Probability and Mathematical Statistics (Continued)

PURI, VILAPLANA, and WERTZ • New Persp
 Applied Statistics
RANDLES and WOLFE • Introduction to the
 Statistics
RAO • Linear Statistical Inference and Its Applic
RAO • Real and Stochastic Analysis
RAO and SEDRANSK • W.G. Cochran's Impact on Statistics
RAO • Asymptotic Theory of Statistical Inference
ROHATGI • An Introduction to Probability Theory and Mathematical Statistics
ROHATGI • Statistical Inference
ROSS • Stochastic Processes
RUBINSTEIN • Simulation and The Monte Carlo Method
SCHEFFE • The Analysis of Variance
SEBER • Linear Regression Analysis
SEBER • Multivariate Observations
SEN • Sequential Nonparametrics: Invariance Principles and Statistical Inference
SERFLING • Approximation Theorems of Mathematical Statistics
SHORACK and WELLNER • Empirical Processes with Applications to Statistics
TJUR • Probability Based on Radon Measures

Applied Probability and Statistics

ABRAHAM and LEDOLTER • Statistical Methods for Forecasting
AGRESTI • Analysis of Ordinal Categorical Data
AICKIN • Linear Statistical Analysis of Discrete Data
ANDERSON, AUQUIER, HAUCK, OAKES, VANDAELE, and WEISBERG • Statistical Methods for Comparative Studies
ARTHANARI and DODGE • Mathematical Programming in Statistics
ASMUSSEN • Applied Probability and Queues
BAILEY • The Elements of Stochastic Processes with Applications to the Natural Sciences
BAILEY • Mathematics, Statistics and Systems for Health
BARNETT • Interpreting Multivariate Data
BARNETT and LEWIS • Outliers in Statistical Data, *Second Edition*
BARTHOLOMEW • Stochastic Models for Social Processes, *Third Edition*
BARTHOLOMEW and FORBES • Statistical Techniques for Manpower Planning
BECK and ARNOLD • Parameter Estimation in Engineering and Science
BELSLEY, KUH, and WELSCH • Regression Diagnostics: Identifying Influential Data and Sources of Collinearity
BHAT • Elements of Applied Stochastic Processes, *Second Edition*
BLOOMFIELD • Fourier Analysis of Time Series: An Introduction
BOX • R. A. Fisher, The Life of a Scientist
BOX and DRAPER • Empirical Model-Building and Response Surfaces
BOX and DRAPER • Evolutionary Operation: A Statistical Method for Process Improvement
BOX, HUNTER, and HUNTER • Statistics for Experimenters: An Introduction to Design, Data Analysis, and Model Building
BROWN and HOLLANDER • Statistics: A Biomedical Introduction
BUNKE and BUNKE • Statistical Inference in Linear Models, Volume I
CHAMBERS • Computational Methods for Data Analysis
CHATTERJEE and PRICE • Regression Analysis by Example
CHOW • Econometric Analysis by Control Methods
CLARKE and DISNEY • Probability and Random Processes: A First Course with Applications, *Second Edition*
COCHRAN • Sampling Techniques, *Third Edition*
COCHRAN and COX • Experimental Designs, *Second Edition*
CONOVER • Practical Nonparametric Statistics, *Second Edition*
CONOVER and IMAN • Introduction to Modern Business Statistics
CORNELL • Experiments with Mixtures: Designs, Models and The Analysis of Mixture Data

Applied Probability and Statistics (Continued)

COX • Planning of Experiments
COX • A Handbook of Introductory Statistical Methods
DANIEL • Biostatistics: A Foundation for Analysis in the Health Sciences, *Fourth Edition*
DANIEL • Applications of Statistics to Industrial Experimentation
DANIEL and WOOD • Fitting Equations to Data: Computer Analysis of Multifactor Data, *Second Edition*
DAVID • Order Statistics, *Second Edition*
DAVISON • Multidimensional Scaling
DEGROOT, FIENBERG and KADANE • Statistics and the Law
DEMING • Sample Design in Business Research
DILLON and GOLDSTEIN • Multivariate Analysis: Methods and Applications
DODGE • Analysis of Experiments with Missing Data
DODGE and ROMIG • Sampling Inspection Tables, *Second Edition*
DOWDY and WEARDEN • Statistics for Research
DRAPER and SMITH • Applied Regression Analysis, *Second Edition*
DUNN • Basic Statistics: A Primer for the Biomedical Sciences, *Second Edition*
DUNN and CLARK • Applied Statistics: Analysis of Variance and Regression
ELANDT-JOHNSON and JOHNSON • Survival Models and Data Analysis
FLEISS • Statistical Methods for Rates and Proportions, *Second Edition*
FLEISS • The Design and Analysis of Clinical Experiments
FOX • Linear Statistical Models and Related Methods
FRANKEN, KÖNIG, ARNDT, and SCHMIDT • Queues and Point Processes
GALLANT • Nonlinear Statistical Models
GIBBONS, OLKIN, and SOBEL • Selecting and Ordering Populations: A New Statistical Methodology
GNANADESIKAN • Methods for Statistical Data Analysis of Multivariate Observations
GREENBERG and WEBSTER • Advanced Econometrics: A Bridge to the Literature
GROSS and HARRIS • Fundamentals of Queueing Theory, *Second Edition*
GUPTA and PANCHAPAKESAN • Multiple Decision Procedures: Theory and Methodology of Selecting and Ranking Populations
GUTTMAN, WILKS, and HUNTER • Introductory Engineering Statistics, *Third Edition*
HAHN and SHAPIRO • Statistical Models in Engineering
HALD • Statistical Tables and Formulas
HALD • Statistical Theory with Engineering Applications
HAND • Discrimination and Classification
HOAGLIN, MOSTELLER and TUKEY • Exploring Data Tables, Trends and Shapes
HOAGLIN, MOSTELLER, and TUKEY • Understanding Robust and Exploratory Data Analysis
HOEL • Elementary Statistics, *Fourth Edition*
HOEL and JESSEN • Basic Statistics for Business and Economics, *Third Edition*
HOGG and KLUGMAN • Loss Distributions
HOLLANDER and WOLFE • Nonparametric Statistical Methods
IMAN and CONOVER • Modern Business Statistics
JAGERS • Branching Processes with Biological Applications
JESSEN • Statistical Survey Techniques
JOHNSON • Multivariate Statistical Simulation
JOHNSON and KOTZ • Distributions in Statistics
Discrete Distributions
Continuous Univariate Distributions—1
Continuous Univariate Distributions—2
Continuous Multivariate Distributions

(*continued on back*)

Statistical Design
for Research

Statistical Design For Research

LESLIE KISH

Research Scientist, Institute for Social Research
Professor Emeritus, Department of Sociology
The University of Michigan

JOHN WILEY & SONS

New York • Chichester • Brisbane • Toronto • Singapore

Library of Congress Cataloging-in-Publication Data:

Kish, Leslie, 1910–
Statistical design for research.

(Wiley series in probability and mathematical statistics. Applied probability and statistics section, ISSN 0271–6356)
Bibliography: p.
Includes index.
1. Experimental design. 2. Social sciences—Statistical methods. I. Title. II. Series: Wiley series in probability and mathematical statistics. Applied probability and statistics.

QA279.K54 1987 001.4′22 86-28084
ISBN 0-471-08359-3

Printed in the United States of America

10 9 8 7 6 5 4 3 2

Preface

In this book I address some basic aspects of research design that are central and common to many related fields. They are central to the social sciences of economics, sociology, psychology, and political science and, more broadly, to health sciences, research in education and social welfare, and evaluation research in medical and environmental fields. Market research and industrial research workers should also be concerned with these problems, as should researchers in agricultural economics and production. Admittedly, these aspects and problems are less central to laboratory research in physics and chemistry and to fields like astronomy. But they are central and common to most kinds of research concerned with humans.

A common core of problems of *statistical design* exists in all these fields, along with many basic similarities in their feasible solutions. Hence a unified presentation seems more economical than separate treatments for each field would be; such unification also gains additional heuristic strength from many fruitful analogies and from the "portability" across fields from examples of both successes and failures.

Among all these fields there exist substantial differences in theory, in measurement methods, and in many practical and nonstatistical aspects of research designs. All those aspects are better left to separate, specialized treatments for each field. But the *statistical* aspects of design can be better perceived and presented within a common framework. That separation of the statistical from the other aspects of design defines one boundary for the aims of this book.

Another boundary is imposed by my aim to deal with statistical *design*, but not with statistical analysis; and this distinction explains my selection and treatment of topics. Most of statistics—courses, books, and journals—deal primarily or only with analysis and only rarely and little with design; they deal with estimation and computation and not with selection and collection. "Given *n* random cases of some variable" are typical starting words for statistical fables, and each word is misleading for real data. Most data are not "given"—they have to be taken, enticed, captured, or mined. The *n* is generally not fixed but varies, because of many imperfections in collection. The selection is not simple "random," but clustered and stratified or otherwise complex. Also one obtains not true "variables," but only observations subject to errors, which the analyst should recognize and control. Such problems are all treated in statistical design.

Statistical design concerns aspects and problems that belong to statistics and to statisticians, because it and we are (or should be) best equipped to deal with them. They are largely omitted from statistical analysis, which is organized closely around a mathematical core. The aims and contents of this book concern the methods and philosophy of statistics, but they are mostly nonmathematical, and that may be why they are largely neglected in the statistical literature.

I suggest that from the ill-defined, broad, and general area of statistical design, two well-defined and specialized approaches have been carved out and then refined and shaped into mathematical disciplines. Books on modern *experimental designs* began with Fisher [1935], but they all deal almost entirely with symmetrical designs for pure experiments. Books on modern *survey sampling* begin with Yates [1949 and 1981], but such books all deal almost entirely with descriptive statistics only. (Of course, both fields acknowledge earlier ancestors.) Knowledge gained from both fields—gained by both me and you the readers—is useful for the ideas advanced in this book. But between them both disciplines fail to cover the primary and basic aspects of statistical design, so vital in research, that are the subjects of this book.

We need several more sources of methods to encompass the area of statistical designs. The most important sources come from the broad literature on *observational studies,* i.e., *controlled investigations*, inquiries, clinical trials, quasi-experimental designs, etc., diffused in various fields of applications, like epidemiology, educational psychology, sociological research, and recently evaluation research. The authors try to impart their accumulated wisdom in teachable capsules, a difficult task because the topics are diffuse, compared with the disciplines for the designs of experiments and of samples. Such efforts must be made, but only a statistical framework can capture the statistical aspects of designing observational studies (Chapter 3).

How do I hope to add to the literature already available in the three disjoint fields? First, I point out and emphasize that choosing among the three methods should be a primary statistical decision, based on the three criteria of "Representation, Randomization, and Realism" (Chapter 1). Second, by joining together knowledge from the three fields, we build strength from the statistical unity that exists in the three methods. In particular, "Analytical Uses of Sample Surveys" (Chapter 2) provides tools for the understanding and treatment of observational studies also. Third, both the similarities and differences between true experiments and observational studies are clarified with a system of modules for sources of bias in "Designs for Comparisons" (Chapter 3). This is followed by descriptions of designs for "Controls for Disturbing Vairables" (Chapter 4). I have tried out these approaches in teaching and used them in practice in designing, consulting, and writing research articles. This is also true of Chapters 5, 6, and 7.

Frankly, I worked hard and long to make this novel book useful not only for graduate courses, but also for the reference shelves of researchers, as tools for their continuing education, and useful not only for the designers of research, but also for consumers of their research designs. Using this book should promote mutual comprehension between consultants and consultees. "The statistician cannot excuse himself from the duty of getting his head clear on the principles of scientific inference, but equally no other thinking man can avoid a like obligation" [Fisher 1935].

I wish to note here several features, some of them unusual, designed for the readers' convenience. The table of contents, with the titles of chapters, sections, and subsections, is full and long. It is followed by a novel "chapter and section contents," which presents sentence summaries for all sections. This should help readers find topics for reference and help cross-referencing. At the back of the book there is a subject index with references to sections. But instead of an author index, I have included something I hope is more useful: The long listing of references shows for each publication the sections(s) of this book where it was cited. Thus the reader can readily find the context of its use and its relation to this book. I have included several problems at the end of each chapter. Also tables and figures, most of them new, illustrate many of the central aspects, and the figures carry rather full explanatory legends.

The book is organized to facilitate its use for consulting as well as for courses in methods and statistics. Within the chapters, sections and even subsections can be read separately for specific research problems. For these purposes I include some repetitions and many cross-references to related sections, often in other chapters.

For courses, each of the chapters can be used separately and in any order, not just successively. They are organized internally, but with many references to other sections and to outside sources. Chapter 1, on the relations between surveys, experiments, and observational studies, can supplement courses in all three fields, also in statistical methods. Chapter 2, on "Analytical Uses of Sample Surveys," can complement various courses in sampling, experiments, or statistical design. Chapters 3, 4, 5 and 6—on "Designs for Comparisons," "Controls for Disturbing Variables," and "Sample Designs over Time"—can fit into various courses on methods. Chapter 5, on "Samples and Censuses," complements courses on sampling. On the other hand, this book can also serve as the core for a course on methods, to be supplemented with books on experimental designs, observational studies, and other methods.

Many other topics of design are fitted into appropriate sections of Chapters 1–6 and into the six miscellaneous sections of Chapter 7, "Several Distinct Problems of Design." These topics should be sought in the table of contents and the index before readers decide that their favorite subject has been omitted or treated too briefly. But this may happen. Most obvious is the lack of treatment for experimental designs, despite many references and uses of its technical terms. Many good books that deal solely with the subject are available, all longer than this volume; the topic also appears prominently in other books and courses; thus a brief consideration here did not seem worthwhile. Survey sampling can also be explored elsewhere, but its analytical uses, often neglected, are treated briefly in Chapter 2 and in Section 7.1. It was more difficult for me to omit statistical estimation, because that is so intimately bound to selection and design. Nevertheless I had to omit it, because there is no separating of estimation from analysis and computing, and they occupy the vast bulk of statistical learning, of which readers must have had modest shares, at least. However, a few simple estimation methods are included in Sections 4.5, on standardization; 4.7, on ratio estimates; and 7.4, on weighted estimates, because they are often useful yet missing from most current treatments. A glossary of terms, always sought after, did not seem feasible within reasonable length, because three distinct fields of methods for research designs were covered, plus related topics.

I felt compelled to condense too many ideas on design into a short book, so that they will be read, not only written and included. Hence this book may seem very "dense" in ideas per page, as two readers remarked. I urge readers to persevere; some sections will become clearer on a second or even third reading. Some of the contents and ideas will be more novel and controversial than students ordinarily encounter in textbooks. Therefore this book needs brave and good teachers, prepared with advanced reading of collateral and contradictory references, to deal with controversies. But I hope that such controversies will be enlightening as well as stimulating and heated.

Earlier I wanted to provide more than one side of controversial topics, and such attempts may interest readers in Sections 1.5, on statistical tests; 1.7, on representation and probability sampling; 1.8 on model-dependent inference; 2.7, on analytical studies; 3.1, on substitutes for probability sampling; 3.5, on external/internal validity; and 6.1A, on longitudinal studies. But it would more than double the pages to do full justice to two sides on all issues. But most controversies have x sides, $x > 2$ and variable! The readers, students and teachers must supply their own $y < x - 1$ sides! I tried to signal the more controversial items, often by emphasizing the personal *I*.

In summary, I attempt in this book to claim for statistics the primary and most vital aspects of research design. Too often problems reach consulting statisticians far too late, after the data have been collected with inefficient or faulty designs. But it is also too late, even before collection, if the statistician is merely asked to select a sample of n households for one city; or, say, for two pairs of treatment plus control cities. By then the primary decisions of statistical design will already have been made: method of design, type and number of primary units (sites), and number and kind of cases. The crucial aspects of statistical design would have been decided, without the valuable improvements that statisticians may contribute to the validity and efficiency of the design. The chief aim of this book is to bring statistics into those early phases of decision making and design. It could be claimed that these topics belong to statistical or scientific philosophy. But I have not noted a rush of philosophers into the area.

I have taught, lectured, and written on these topics and for about 13 years hesitatingly worked on this book, while collecting more experience and material. The vastness and vagueness of the area inhibited my efforts, but I was impelled by the need I felt and by the urgings of friends and colleagues. Among them I single out two, most often cited here, William Cochran and Donald Campbell, who "egged" me on. My wife, Rhea, not only edited it, but also made me "live to see the Day." I was fortunate to have the patience, confidence, and support of my editor Beatrice Shube. Katherine Metcalf did most of the typing gracefully and skillfully through $x + 1$ versions of each section. I am grateful to all of them.

LESLIE KISH

Ann Arbor, Michigan
January 1987

Contents

Chapter and Section Contents

These brief notes should help readers find their way into chapters and sections, when the table of contents seems not informative enough. They may be useful both before and after reading the chapters. However they hardly serve as adequate summaries of the contents, because these are too varied, brief, and dense for condensation.

1. Representation, Randomization and Realism

(1) Choice and compromise among the three preceding criteria should be the primary act of research design. (2) In addition to two sets of Explanatory variables, the Predictors X and Predictands Y, three classes of Extraneous variables are defined: Controlled C, Disturbing D, and Randomized R. (3) Surveys, experiments, and controlled investigations denote three major types of designs that excel respectively in each of the three criteria, but neglect the other two. In "ideal" experiments all D variables are removed into either C or R. (4) Both randomization over treatments and representation by randomization over populations are proposed as due to the same basic problem: relations between the Explanatory variables X and Y are always conditional on the elements. (5) Statistical tests should be used to separate genuine (population) relations from sampling (chance) variations, and not to measure the relations. (6) An ordered list distinguishes ten designs from descriptive surveys to confined but true experiments. (7) Probability sampling is shown as the prime tool for the broad aim of representation. (8) Population-bound inference based on probability sampling is compared with model-dependent inference.

2. Analytical Uses of Sample Surveys

(1) Four population levels are distinguished by gaps of response and coverage: sampled, frame, target, and inferential. (2) Design effects (deft^2) of complex samples for variances of the means of samples and subclasses are discussed. (3) Sample subclasses represent population domains; they are distinguished between design classes and crossclasses, and between major, minor, and mini classes. (4)

Crossclass means and their differences behave regularly: their design effects become reduced in proportion to crossclass size from the deft^2 for the total sample. (3) Proportionate stratified element sampling (pres) tends to reduce variances only slightly, and much less for crossclasses. (6) Clustering increases variances, sometimes with large deft^2. This $\text{deft}^2 = [1 + \text{roh}\ (\bar{b} - 1)] > 1$ depends on the cluster size $\bar{b}$ and on roh, but $\bar{b}$ and deft^2 decrease in crossclasses. (7) Obstacles to representation for analytical studies are practical and economic, ethical and social, mathematical and historical, and only partly philosophical.

3. Designs for Comparisons

(1) Restricted sites and community studies are substituted for widespread, national samples when these seem too expensive, especially for longitudinal studies. Several divergent sites can provide internal replication. (2) Principal features and limitations motivate the proposed structure of four basic modules for research. (3) The four modules are compared with *relative* costs and variances suitable for any sizes n and with four major types of bias. (4) Five basic designs are compared: one-shot case study, one-group pre/post, control group comparisons, pre/post control groups, and four-group designs. (5) 22 sources of bias, based on Campbell's 12 threats to validity, are classified into six (the four plus two) major types. (6) Responses may be delayed and varied over time, thus requiring more than simple, single, before–after tests. (7) Evaluation research is new and highly specified in objectives: treatment and response, timing and conditions, population.

4. Controls for Disturbing Variables

(1) Control strategies require decisions on loci of control; on selection or analysis; on choice of variables; on numbers of variables and of classes for each. (2) Analysis by separate subclasses is most common and simple, but lacks statistical aggregation. (3) Case-by-case matching is an extreme control by selection, commonly used when control cases are plentiful and accessible. (4) Matching subclasses preserves more cases than case-by-case but loosens controls, and several compromises exist. (5) Standardization preserves all cases with control in analysis, with weights chosen to reduce biases and variances. Indexes denote relative standards. (6) Control of disturbing variables may sometimes be done by complex analysis with covariance, residuals, or categorical data analysis. (7) Ratio estimates are often used to adjust for differences and for errors in data, and are also ubiquitous in sample survey data.

5. Samples and Censuses

(1) Researchers from the outside interact with the Census as users, advisors, etc. (2) Censuses provide precision for domains and local area estimates, but samples are more timely and richer in variables. (3) Samples attached to censuses can yield supplements of richer, better data; checks and evaluations; samples for

separate analysis; auxiliary data for samples. Censuses, samples, and registers can jointly yield postcensal estimates for local areas and small domains.

6. Sample Designs over Time

(1) Alternative possibilities for sampling reference dates, for collecting data, and for reporting periods exist and need more accuracy in terminology. (2) Five purposes (current levels, cumulations, mean changes, individual changes, trends) are matched against designs (partial overlaps, nonoverlaps, complete overlaps, panels, combinations), distinguished by the amount and kind of sample overlaps, which affect variances. (3) Population changes may be internal, or come from external events, or be due to migration, or to boundary changes; each can disturb measures of change. (4) Comparing panels to distinct periodic samples results in entirely different cost and accuracy conclusions than comparing them to retrospective studies. (5) "Split-panel designs" are proposed to combine panels with periodic distinct samples. (6) Similar data from repeated studies can be pooled either by combining summary statistics (e.g., means) or by cumulating cases.

7. Several Distinct Problems of Design

(1) For mathematically intractable sampling errors, several alternatives are described; methods of repeated replications or jackknife estimators and error models seem most useful. (2) Alternative measures for comparisons are possible, and the flexibility of the four basic modules allows extensions. (3) Multipurpose surveys lead to conflicts of design; compromises and "optimal" solutions can be sought and found. (4) Weighted means are shown as a simple common framework for a good variety of selection design problems. (5) Natural organizations come in highly variable sizes, and representative sampling has interesting aspects. (6) "Falsifiability" is a needed philosophical framework for a large variety of design decisions.

Tables and Figures

Statistical Design for Research

CHAPTER 1

Representation, Randomization, and Realism

The statistician cannot excuse himself from the duty of getting his head clear on the principles of scientific inference, but equally no other thinking man can avoid a like obligation. R A Fisher, Design of Experiments.

The difference between cultures is first of all the difference in the questions permitted. V V Nalimov, Faces of Science.

1.1 THREE CRITERIA

Statistical designs always involve compromises between the desirable and the possible. We face inevitable compromises in the choice of the very nature and the structure of statistical designs; in their scope and breadth; and in the size of research projects. Here at the outset we describe the basic compromises in the choice of the nature and structure of the research designs. We also need to justify and to distinguish the designs for observational studies from the two better-known and better-defined fields of survey sampling and of designs for experiments.

The compromises involved in choosing among these three major types of designs concern the basic philosophical problems of all empirical sciences: how to make inferences to large populations, to infinite universes, and to causal systems from limited samples of observations, which are also subject to diverse errors and to random fluctuations. We cannot avoid the basic philosophical problems posed by David Hume in 1740 and restated by other philosophers like Popper [1959], Salmon [1967], and Burks [1977]; and by statisticians like Fisher [1935], Neyman [1934], and others (7.6).

Those problems and compromises always exist implicitly, and sometimes they may even appear vaguely in the justifications for using one of the three major types of designs. I propose, however, that they should be treated explicitly in the design of any empirical research. An example may be helpful here at the outset. Consider designs for comparing and evaluating the relative effectiveness of two or more techniques of instruction in schools. First, compromises are generally needed in the *realism* for the chosen "explanatory" variables: the predictor and the predictand variables, i.e., the treatment and response variables. The different instructional treatments to be compared need to be defined and operationalized; and these tasks will pose difficulties and compromises between ideal and feasible techniques. We may begin with some conceptualized theoretical underlying variables, but then we want to come nearer to the practical realities of eventual and actual applications. Meanwhile we may also need to hew closer to the immediate possibilities of the experimental situation. We must operationalize and control the teaching techniques of the predictor treatments: secure appropriate materials and adequate psychological settings for them and perhaps also recruit satisfactory teachers. However, even greater difficulties and compromises often arise in designating and operationalizing the predictand variables: the criteria for effectiveness of responses. Must we accept simple class tests, or even mere opinions and attitudes of teachers or of students? Or should we try to measure success in subsequent classes? Or can we possibly aim at assessing eventual success in "real" lifetime achievement?

Second, for scientific reliability we should want *randomization* of treatments for subjects, and often this involves great difficulties. It may call for great ingenuity and frequently also for compromises. Can we undertake true randomization of individual students? Or, more likely, classrooms of students? Or must we use entire schools as units of selection? Further, can we obtain rigorous controls for the treatments and uncontaminated measurements for the responses?

Third, the *representativeness* of the sampling units usually involves compromises both in the designation of the target population and in the selection of sampling units from it. What ages and types of students should be designated? Should we aim at a national sample or must we be satisfied with a smaller area, a county or a city? Or should we try only for several contrasting areas, or, contrariwise, should be aim at international comparisons?

This example illustrates the difficulties of satisfying the three criteria of research design to be discussed: realism, randomization, and representation. Yet those difficulties are even greater in many tasks facing social research, such as evaluating the long-range effects of poverty or of welfare policies. Severe compromises must often be made in how we apply the three criteria and in how we choose between the criteria. Placing primary emphasis on realism, on randomization, or on representation tends to result, respectively, in controlled observational studies, in experiments, or in sample surveys. These choices have merely been assumed implicitly (hence largely hidden) both in practice and in theoretical treatments, but they are made explicit in Section 1.3 in order to clarify the sacrifices involved in the compromises.

Choices between designs are facilitated by first discussing in Section 1.2 four classes of variables involved in empirical research. These serve to relate the three major criteria just discussed to the three major types of statistical design defined in Section 1.3.

Within each of these basic categories—criteria, variables, and designs—I aim for clear definitions with one distinct term serving each major type. I chose these terms carefully from good statistical usage. However, we lack uniform terminology in statistics, as in philosophy, and especially between the sciences of economics, sociology, psychology, health, etc., where they are needed and used. I must ask readers to join me in a common vocabulary while reading this book, whatever their personal preferences may otherwise be.

I had less than complete success, because common usage forced me to make a few terms serve double meanings, usually a specific meaning plus a distinct and general usage. *Comparison* refers to the difference of two means (Ch. 3), but also serves in a more general sense, such as in comparisons of costs or efficiencies. *Control* appears in the comparisons of treatment versus control, but also in the more general sense of control of disturbing variables (Ch. 4). *Treatment* appears in those comparisons, usually for a new treatment, whereas *control* denotes standard and older treatments. But treatment/response also denotes predictor/predictand. Of course, we also refer often to treatment of a topic, for example, to denote attention and discussion.

1.2 FOUR CLASSES OF VARIABLES

(E) The *explanatory variables* are the objects of the research design. They are sometimes called *experimental* variables, whereas *explanatory* has sometimes been used for what I shall denote as predictor variables. Here *explanatory* denotes those variables that embody the aims of the research design, among which the researcher wishes to find and measure some specified relationships. The explanatory variables comprise two distinguishable sets: the *predictor* (X) variables comprise the sought causes of the relationships and the *predictand* (Y) variables describe the predicted effects. Other name have been used for these two sets of variables: independent and dependent, stimulus and response, treatment and criteria, cause and effect, X and Y—as well as other pairings of these terms. Determinants and consequences have also been used, especially in demographic research. We are not obliged to explore here the subtle philosophical distinctions between these pairs of terms. The terminology lacks uniformity and I shall opt for neutral terms. Furthermore, in observational studies and in surveys, the differentiation of explanatory variables into predictors and predictands may be more or less arbitrary.

The explanatory variables that comprise the aims of research are designated first on the basis of substantive, scientific theories; they arise from knowledge of and insight into the field under study. Inevitably, however, the potential existence of other, *extraneous* sources of variation must also be recognized. Then methods must be devised for separating these extraneous variables from the explanatory variables. Sorting all the diverse extraneous variables into three classes seems a useful simplification. Furthermore, no confusion need arise from talking here about "variables" instead of "sources" of variation. *The explanatory variables* (E), *predictors* (X), *and predictands* (Y), *embody the aims of the research, and the other three classes of variables are extraneous to those aims* (see Figure 1.2.1).

(C) *Controlled variables* comprise those extraneous variables that can be controlled adequately by the research design. Control may be exercised either by design of the selection procedures or by estimation techniques in the statistical analysis, or perhaps by both. The choices depend on foresight and knowledge, but also on the availability and strategical use of data and of resources. Techniques for controlling extraneous variables are aimed at decreasing random errors (class R), or decreasing the biasing effects of disturbing variables (class D), or both.

(D) *Disturbing variables* are uncontrolled extraneous variables, which may be confounded with the explanatory variables (class E). Failure to remove all of these D variables either into class C of controlled variables or into class R of randomized variables is the primary disadvantage and concern of nonexperimental designs.

(R) *Randomized variables* are uncontrolled extraneous variables that are treated as random errors. In "ideal" experiments they are actually, operationally randomized, but in surveys and investigations they are only assumed to be randomized—as discussed in 1.4. Randomization may be viewed as a form of experimental control, but distinct from the forms used for class C variables.

With *efficient* designs we aim to place into class C (controlled variables) as much of the extraneous variables as seems feasible, practical, and economi-

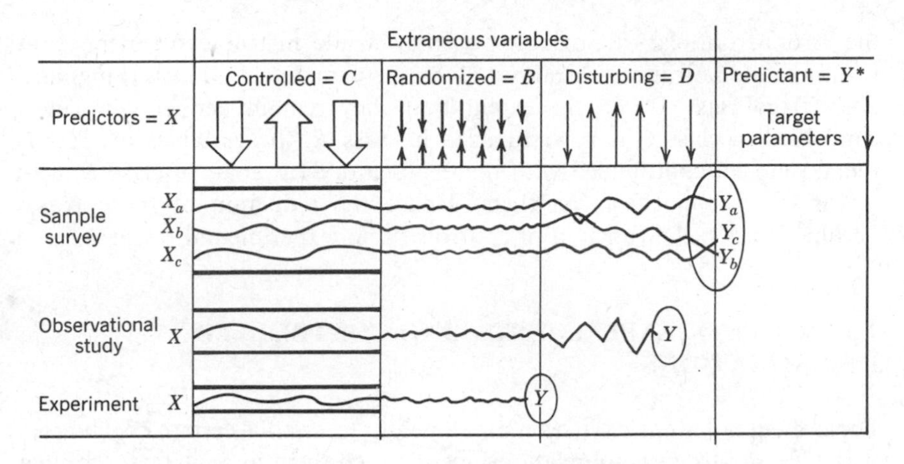

Figure 1.2.1. Effects of three classes (C, R, D) of extraneous variables on the explanatory (E) variables ($X \rightarrow Y$).

Three types of extraneous variables may disturb the path of the explanatory (E) variables, from predictor (X) to predicand (Y). Some variables are controlled (C) with various kinds of controls (blocking, stratification, matching weighting, etc.). Others are randomized (R) in experiments, or treated as randomized in surveys and observational studies. However, some effects remain as disturbing (D) variables in surveys and in observational studies, but none of these remain in "true" experiments.

Arrows going up and down signify the conflicting effects of extraneous variables. Controlled variables are shown thicker than disturbing variables, in the hope that the strongest variables get controlled by the design. The randomized variables are shown as numerous, weak, and acting in both directions, tending to cancel; their effects on the ($X \rightarrow Y$) paths are shown as relatively weak.

Experiments are shown as most tightly controlled; and sample surveys as much less so; but they can represent broader population bases that cover several domains, as shown by the three ($X \rightarrow Y$) paths, a, b, c.

However, the predictands Y of experiments are shown as much further from the target parameter Y^* that represents the ultimate response/effect the study attempts to predict/estimate. Those gaps refer to the relative lacks of "realism" of study designs.

cal. However, it is not possible or practical to control more than a few of the potentially disturbing variables, and some, or most, must be left uncontrolled. The aim of randomization in experiments is to place all class D variables into class R. In the "ideal" experiment there are no variables left in class D; all extraneous variables have been either controlled in class C or randomized in class R. By placing disturbing variables into class C we eliminate the effects they would have in class R. Though random errors are left in class R, biases due to class D are eliminated in "ideal" experiments.

However, in nonexperimental research—in surveys and in investigations—controls must do double duty. They increase efficiency by reducing the errors from class R variables, as they would in true experiments. But without randomization we cannot eliminate completely all disturbing variables from class D. To the extent that they remain neither completely controlled in class C nor randomized in class R, the variables of class D remain mixed, confounded with the explanatory variables of class E, with unknown biasing effects on them. Hence the reduction of those biases becomes the crucial function of controls in nonexperimental research.

1.3 SURVEYS, EXPERIMENTS, AND CONTROLLED INVESTIGATIONS

Experimental designs have been developed to test and ascertain explanatory variables and to measure relationships between them in analytical probings of data. Separately and distinctly, the theory of survey sampling has been developed chiefly to provide descriptive statistics, means, proportions, and totals—particularly for large samples from much larger populations. However, survey data are used frequently and successfully, especially in the social sciences, for statistical and analytical research into relationships among explanatory variables. Furthermore, neither experiments nor sample surveys are feasible or practical in many situations, and controlled investigations and observational studies of some kind must be often used. The theoretical development of each of these three types of designs has been made in splendid isolation, in order to simplify and clarify them. The connections and contrasts between them have not been much explored, but we are about to embark on that exploration.

By *experiments* I mean here "ideal" experiments in which all the extraneous variables have been either controlled or randomized (1.4). By *surveys* (or *sample surveys*), I mean probability samples in which all members of a defined population have a known positive probability of selection into the sample (1.7). By *investigations* (or *controlled investigations*), I mean the collection of data—with care, and often with considerable control—without either the randomization of experiments or the probability sampling of surveys (Ch. 3). The differences between experiments, surveys, and investigations are not the consequences of statistical analysis, which may be similar; they result from different designs for introducing the variables and for selecting the population elements (subjects).

In considering the larger ends of any scientific research, only part of the total means (i.e., resources) required for inference can be brought under objective and firm control; other parts must be left to more or less vague and subjective—however skillful—judgment. Scientists seek to maximize the first part and thus to minimize the second. In assessing the ends, the costs, and

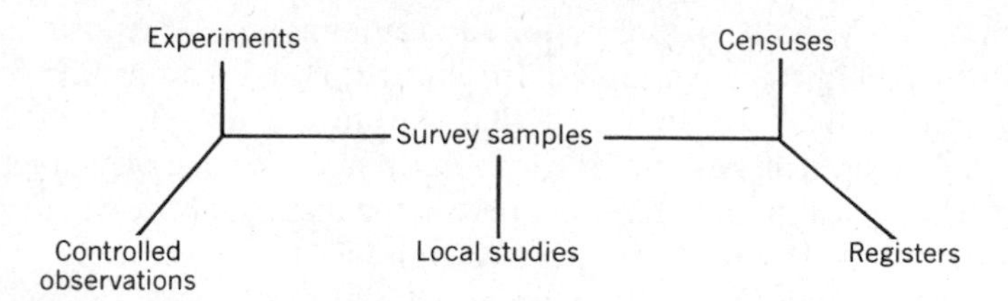

Figure 1.3.1. Three different comparisons of survey samples.

Comparisons of survey samples with experiments and with controlled observations are discussed in this section (1.3). These three methods are justified as preferred strategic choices for the three criteria of representation, randomized treatments, and realism. Another kind of strategic choice involves the relative advantages of sample surveys against censuses or registers (5.2). Still another comparison concerns the choice between widespread sample surveys and local studies confined to one or a few sites (3.1). Thus sampling surveys are shown as alternatives to three distinctly different methods of data collection, and that is the only reason for their central position in the figure.

the feasible means they make a strategic choice of methods. Thus they are faced with the three basic problems of scientific research: randomization, representation, and the realism of measurements.

Experiments are strong on control of the explanatory variables through the randomization of predictor variables over subjects (i.e., subjects over treatments); but they are weak on representation over defined target populations, and often also on the realism of measurements. Surveys are strong on representation, but they are weak on control of variables. Investigations are weak on control and often on representation; their great prevalence is due often to low cost and relative convenience, and at times to the need for and feasibility of realism of measurements in "natural settings." We are faced usually with conflicts between desires for randomization, for representation, and for realism. It is seldom that desires for all three criteria can be satisfied adequately in one research design, and very often desires cannot be satisfied for even two of the three. More often people merely emphasize one criterion because it is least costly and most convenient, and because it may appear on theoretical grounds—convincingly or only hopefully—the most justified. But when hope or wish is the father of the thought, the offspring may be illegitimate. The criteria that must be sacrificed must be considered more thoroughly. We ask for a greater role for explicit models of the diverse sources of variation that arise in the three types of research designs (see Figure 1.3.1).

Experiments have three chief advantages: (1) Through randomization of extraneous variables the biases from disturbing variables (class *D*) can be eliminated. (2) Controls over the introduction and variation of the predictor variables clarify both the direction and the nature of causation from predictor to predictand variables. In contrast, for relationships found in survey results, that direction and that nature are not clear. (3) Modern designs of experiments allow for great flexibility, efficiency, symmetry, and powerful statistical manipulation, whereas the analytical use of survey data presents difficult statistical problems (2.7 and 7.1).

The advantages of the experimental method are so well known that we need not dwell on them here. It is considered the scientific method par excellence—when feasible. In many situations experiments are not feasible, and this is often the case in the social sciences, because predictor variables cannot be assigned freely to subjects for several reasons. (1) It is impossible to assign predictors like age, sex, religion, I.Q., and even income and education in most cases. (2) It is unethical to assign predictors like smoking, drug abuse, and even a new medicine or placebo in some situations. (3) It is impractical to assign a new teaching method to separate students within classes, or management styles to separate employees within workgroups, etc. Yet it is a mistake to use these situations to separate the social from the physical and biological sciences. Such situations also occur frequently in the physical sciences (in meteorology, astronomy, geology), in the biological sciences, and in medicine, engineering, manufacturing, business, etc. (1.4).

Even where feasible, the experimental method may also have short-comings that must be either overcome or tolerated. First, often it may be difficult to choose and operate the control variables; it may be difficult or impossible to design an "ideal" experiment. Newspapers bring frequent reports of successes with new drugs and cures that later may be either denied or forgotten. We shall discuss later (Ch. 4) the need for and difficulties with proper controls, and readers have their own ideas and examples. Here we merely mention the use of placebos to hide treatments from subjects and of "double-blind clinical trials" to hide them both from subjects and from researchers (1.4). We shall refer to the famous Hawthorne experiments with workers (3.6) and to the great Pavlov with his dogs.

In actual experiments the "ideal is not reached easily, hence the distinction between all experimental and nonexperimental research is not absolute. Troubles with experimental controls misled even the great Pavlov into believing temporarily that he had proof of the inheritance of an acquired ability to learn: "In an informal statement made at the time of the Thirteenth International Physiological Congress, Boston, August, 1929, Pavlov explained that in checking these experiments it was found that the apparent improvement in the ability to learn on the part of successive generations of mice was really due to an improvement in the ability to teach on the part of the experimenter" [Greenberg, 1929, p. 327]. Nevertheless the distinction is real and worthwhile. Hence Sir Austin Bradford Hill overstates his good case by saying: "the difficulties of experiments are no less," when discussing the difficulties of observational studies [in Cochran, 1965]. [Kish 1959, 1975]

Thus, the advantages of experiments over surveys, in permitting better control of treatments, are only relative, not absolute. The design of proper experimental controls is not automatic; it is an art requiring scientific knowledge and foresight in planning the experiment, and hindsight in interpreting the results. Nevertheless, the distinction in control between experiments and surveys is real and considerable. To emphasize this distinction we refer here to "ideal" experiments in which the control of biases from disturbing variables is complete.

Second, it is generally difficult to design experiments so as to represent a specified important population. In fact, the questions and problems of sampling, i.e., of the *representation* of specified populations in experimental results, have been largely ignored in experimental design. Both in theory and in practice, experimental research has often neglected the basic truth that causal systems, the distributions of relations—like the distributions of characteristics—exist only within specified universes. The statistical inferences derived from the experimental testing of several treatments are restricted to the population(s) included in the experimental design. However, they must receive much broader applications, and for that purpose we must resort to models either explicitly or—too often—only implicitly. These controversial issues are treated briefly below (1.4 and 1.8).

Third, for many research aims, especially in the social sciences, contriving the desired *realism* of a "natural setting" for the measurements is not feasible in experimental designs. Hence social experiments sometimes give answers to questions that have only vague meanings. That is, artificially contrived experimental variables may have only tenuous relationships to the variables the researcher would like to investigate.

The second and third weaknesses of experiments point to the advantages of surveys. Not only do probability samples permit clear statistical inferences to defined populations, but the measurements can often be made in the "natural settings" of actual populations. Thus in practical research situations the experimental method, like the survey method, has its distinct problems and drawbacks as well as its advantages.

In social research some designs of controlled investigations are frequently preferred over both surveys and experiments and are chosen—for reasons of cost, or feasibility, or the preservation of the desired realism of the measurements. Ingenious adaptations of experimental designs have been contrived for these controlled investigations (Ch. 3). The statistical framework and analysis of experimental designs are often used, but without the randomization of true experiments. Great ingenuity is often needed in these designs to provide flexibility, efficiency, and especially some control over the extraneous variables (Ch. 4). Controlled investigations take many forms and have many names: observational studies, controlled observations, quasi- or pseudo-experiments, natural experiments, and so on. I prefer not to borrow the prestige work *experiment* for studies where the predictors are not randomized, and prefer to avoid terms like *ex-post-facto experiments*, that confuse more than help. Controlled investigations and sample surveys are not merely second-class experiments; they have their own justifications.

In practice we usually lack the resources to overcome all difficulties, and thus to achieve simultaneously the perfection of realism of measurements, of randomization to control treatments, and of representation over large populations. Let us agree that often, even usually, we cannot satisfy all these three criteria simultaneously. After admitting that much, however, many writers proclaim an overall hierarchy among these criteria, so that one criterion is paramount in all situations. For example, some believe that randomization of treatments (or "internal validity"), when possible, must be had at all costs, before considering representation over populations (or "external validity") or realism. On the contrary, I believe that there is no supercriterion that would lead to a unique, overall, and ubiquitous superiority among the three criteria. Rather, one must choose and compromise with a research strategy so as to fit our resources to the situation at hand. In any specific situation one method may be better or more practical than the others; but there is no overall superiority in all situations for any of the three designs. Understanding the advantages and weaknesses of each should lead to better choices. Sometimes, a slight relaxing of one criterion can lead to large gains in one or even in both of the other two criteria. Furthermore, some great research problems need to be attacked separately with two or with all three methods.

These antihierarchical views on strategy are in partial disagreement with some of the best writing on controlling variables in social experiments and investigations by Campbell [1957] and Campbell and Stanley [1963], with which the reader should be acquainted (see end of 3.5).

Each of the three kinds of design can be improved with efforts to overcome their specific weaknesses. Because the chief weakness of surveys is lack of control over treatments, survey researchers should improve their collection and their use of auxiliary variables as controls against disturbing variables. They should become more alert to social changes and use them to measure the effects of "natural experiments." They should more often explore survey data with analytical, multivariate techniques—a trend well under way now.

On the other hand, experiments and controlled investigations can often be improved with efforts to specify their populations more clearly and to make their results more representative of the populations. Often more can and should be done to broaden the design base to facilitate statistical inference to wider and more significant populations. Researchers too often and too early commit themselves to small, convenient, or captive populations. Too often the researcher justifies those restrictions as attempts to make the subjects more "homogeneous." If common sense will not dispel this error, reading R A Fisher may.

> We have seen that the factorial arrangement possesses two advantages over experiments involving only single factors: (i) Greater *efficiency* in that these factors are evaluated with the same precision by means of only a quarter of the number of observations that would otherwise be necessary; and (ii) Greater *comprehensiveness* in that, in addition to the 4 effects of single factors, their 11 possible interactions are evaluated. There is a third advantage which while less obvious than the former two, has an important bearing upon the utility of the experimental results in their practical application. This is that any conclusion, such as that it is advantageous to increase the quantity of a given ingredient, has a wider inductive basis when inferred from an experiment in which the quantities of other ingredients have been varied, than it would have from any amount of experimentation, in which these had been kept strictly constant. The exact standardization of experimental conditions, which is often thoughtlessly advocated as a panacea, always carries with it the real disadvantage that a highly standardized experiment supplies direct information only in respect to the narrow range of conditions achieved by standardization. Standardization, therefore, weakens rather than strengthens our ground for inferring a like result, when, as is invariably the case in practice, these conditions are somewhat varied. [Fisher, 1953, p. 99]

The researcher should view the population base in terms of cost factors and components of variation; then broaden the base of statistical inference as much as the resources allow (3.1 and 7.6).

1.4 RANDOMIZATION OF SUBJECTS OVER TREATMENTS AND OVER POPULATIONS

Under this title I shall discuss the needs for randomization and for representation as two *related* aspects of research design. Even more than related, I propose that those two needs have common theoretical, philosophical roots. To recognize those common roots honestly is important; but it is also difficult, because in practice we usually must make choices between the two needs, sacrificing and compromising between them for economic and strategic reasons, as discussed in Section 1.3.

The needs for randomization of treatments and for representation are usually treated separately. I assume that readers are acquainted with some of the arguments for both, and they will find other support in Chapters 2 and 3. The arguments for randomized treatments in experiments and for probability selections in sample surveys are both strong—and we must recognize them even when we see them so often sacrificed for reasons of cost and feasibility.

That the two kinds of randomization have common theoretical roots is not commonly recognized. But that view is important not only for a clearer philosophical understanding, but also for several practical reasons in the design for most research. First, the very choice between major design types, especially between experiments and surveys, should not be arbitrary; we should consider not only relative feasibilities and costs but also the nature and scope both of gains and of sacrifices in inference with each kind of design. This has been our theme in Sections 1.1–1.3. Second, this view provides the thrust for extending the scope of representation for experiments, so as to facilitate broader inferences from experiments (3.1 and 7.6). Third, this view motivates the analytical uses of sample surveys (2.7 and 7.1).

The practice and theory of survey sampling recognize that it must deal with populations of elements with fixed values Y_i, whose distribution is unknown; that it cannot simply assume its sample values to be "I.I.D.r.v." (independently and identically distributed random variables). Hence we use randomization over the population of elements in order to obtain probability selections of elements into the samples. The values of statistics (such as the sample mean $\bar{y}$ for estimating the population parameter $\bar{Y} = \Sigma Y_i/N$) depend on the selection, with known probabilities, into the sample of some of the N population values $Y_i (i = 1, 2, 3, \ldots, N)$.

Now I propose that the sampling view is philosophically valid and necessary not only for single variables but also for *relations* between two or more variables. Classical models for relations between random variables disregard the necessity for population bases for the physical, biological, or social subjects of the real world. We survey samplers propose a *population-bound* approach instead of the *model-based* classical approach. The terms *design-based*, *randomization*, and *empirical* also appear for the population approach, and *population-free*, *model-dependent*, and *mathematical* can also describe the classical model approach. But I propose the preceding neutral terms, because all approaches must use models and mathematics, and because mathematics and models need not be population free, and I propose they should not be (1.8).

Using a population-bound approach, we note that in the real world the relations between variables are properties of specific subjects that are the elements of defined or definable populations. Although we may imagine, talk, and write about relations between variables, and between random variables, for empirical research those relations must be eventually translated and attached to population elements. Such attachment is necessary at both ends of any empirical research process: first when making observations (whether in surveys or in experiments) and last in making inferences. Only during the intervening mathematical analyses can the linkage be neglected. The linkage is neglected also during abstract discourses about theory that can be both population-free and data-free; and those discourses show vast gaps between statistical theory and practice.

Therefore I suggest we remember *that all relations in the physical world between predictor and predictand variables are conditional on the elements of the population subjected to research.* All relations of stimulus to response are dependent on the subjects involved. That the stimulus–response relation is not constant, not uniform, not deterministic should be obvious in all observations—or we would not need statistics. The inherent variations in those relations result in errors that are not small enough to be negligible. Furthermore, the errors are not purely, simply random—in other words, not independently and identically distributed (I.I.D.)—and that is why we need randomizations both over treatments and over populations.

A modest use of symbols will help here. A value of the ith element in the population of N elements can be written as $Y_i = \bar{Y} + d_i$ around the common mean $\bar{Y} = \Sigma Y_i/N$. This we try to estimate with the sample mean $\bar{y}$ with a standard error, which is a function of the d_i and of the sample design for selecting them. Furthermore, the Y_i may also represent vectors of the many variables measured in most surveys and each with its mean $\bar{Y}$ and its own distribution of deviations d_i.

These deviations have diverse complex distributions not only with regard to the overall frequency curve of the values of d_i but also in their spread within the populations: The populations generally are not "I.I.D.," like a "well-churned urn." Probability selection with randomization over the whole population is the accepted sampling strategy for dealing with these problems, the chief of which concerns the avoidance of large biases due to sample selection (1.7). The effects of diverse complex sample designs is the subject of survey sampling (Ch. 2). Those effects are ubiquitous and widely documented for means. They are also documented for other statistics, such as comparisons and analytical relations (7.1). We now look at their sources.

When a treatment effect T_a or T_b is introduced, it is simple and tempting to assume $Y_{ti} = \bar{Y} + T_t + d_i$, where t is either a or b for two treatments; such assumptions are commonly made implicitly. Simple as it is, this amounts to assuming that the effect T_t of treatment t is constant across elements. But if we believe that the treatment–effect (stimulus–response) relation *may* depend on the element (subject), then we should write $Y_{ti} = \bar{Y} + T_{ti} + d_i$, so that the effect T_{ti} depends on both the treatment t and the element i.

If we wish to write $Y_{ti} = \bar{Y} + T_t + e_{ti}$, where T_t is a mean response to treatment t, then we must remember that the error term e_{ti} may depend on both the treatment t and the element i. Those who would write $Y_{ti} = \bar{Y} + T_t + e_i$ must remember and must warn that the e_i not only depend on the elements, but also are specific to the treatment, or use a_i or b_i for treatments a and b.

The classic model of $Y_t = \bar{Y} + T_t + e$, with one common distribution for the errors e, disregards both the interactions between treatments and elements and the variations among elements. These models are based on strong assumptions; and to counter possible biases, randomization has been introduced into modern experimental designs. *Randomization of subjects over treatments is the strategy for eliminating biases in measuring treatment effects due to selection between the experimental subjects.* This randomization counters biases, by averaging differences between treatments over subjects (e_{ti}), both in basic elemental levels (d_i) and in interactions (T_{ti}) with treatments. However, that strategy and the model deal with "internal validity" within the experimental groups only within the scope of the experiment. That strategy of randomization of subjects *within* the experimental groups fails to deal with the differences of subjects e_{ti} *outside* (beyond) the groups, thus with the "external validity" of experiments (3.5).

It would be philosophically difficult to believe that subject differences e_{ti} *within* experimental groups would not also exist *between* potential groups; and the literature contains evidence of diverse and even conflicting results between experiments. To summarize: *Treatment–effect relations depend on subjects; hence those relations are not randomly (I.D.D.) distributed; hence inferences to populations depend on representation.* Nevertheless, representation through randomization of experiments over populations is a problem that is commonly ignored or avoided. The two kinds of randomization are usually treated separately and sometimes even in opposition to each other. Hence I cannot support the arguments above with references and I must appeal directly to the reader's understanding (1.7, 1.8, 2.7, 3.1).

This subject-dependence of stimulus–response relations has been recognized in biological and social research, though it is often still neglected. Standard mice and rabbits are bred and used in order to control the susceptibility of the subjects of research in experiments on cancer and other diseases. In chemical and physical experiments standardized materials and conditions are used; their purity and precision are important aspects of those experimental sciences. But further inference to the worlds outside the laboratories needs considerable further research in chemical engineering, geology, metallurgy, etc. On the other hand, for social research on human subjects such separation of standard subjects under purified conditions is not feasible, nor would it be useful. (See again R A Fisher in 1.3.)

For realism and completeness we should also add the *environment* (or condition, situation, etc.) as conditioning factors for relations between variables. We would then formulate the conditioning (dependence) of relations on elements (subjects) and on the environment as:

$$\text{(predictors} \rightarrow \text{predictand)}|\text{element, environment}$$

or

$$\text{predictand} = f(\text{predictors}|\text{element, environment}).$$

For brevity the effects of environment may be subsumed under the mantle of predictors, or they may be included in the definition of elements, but we must always remember that the same elements may have different predictor–predictand relations in diverse circumstances. For spectacular examples note that the genes that cause sickle cell anemia had been beneficial in malaria-ridden environments. Also genes that cause diabetes may have been useful to tribes that had to endure frequent periods of hunger. Moreover, growing up to be the size of dinosaurs had survival value for reptiles for 100 million years. In the social sciences examples occur daily of the importance of culture and society on individual responses; the readers can supply them from their own experience or reading.

It may be useful to repeat the above in symbols. The classical formulation may be put either generally or for a linear regression as, respectively,

$$y = f(x_g) + e \quad \text{or} \quad y = b_0 + b_1x_1 + b_2x_2 \cdots b_kx_k + e. \quad (1)$$

Here e is the term for errors, and they are assumed to be independently and identically distributed (I.I.D.), with perhaps a known, usually normal, distribution. [Without the error term $y = f(x)$ with $e = 0$, and we would have a deterministic model without need for statistics.] Against population-free models in (1) we propose population-based models, with the subscript i denoting individuals, as

$$y_i = f(x_g) + e_i \quad \text{and} \quad y_i = b_0 + b_1x_1 + b_2x_2 \cdots b_kx_k + e_i, \quad (2)$$

with $e_i(i = 1, 2, \ldots, N)$ being distributed over the N values of the population elements. This extends the earlier simple case for single variables when $y_i = b_0 + e_i$ or $Y_i = \bar{Y} + e_i$. The distribution of the e_i, not completely known, is over a specified population. The definition of the population includes the environment, and the values of the b_g are conditional on it.

The values of the e_i depend on the elements, and accepting this limitation leads to the necessity for randomization of subjects both over treatments (predictors) and over the population of elements. Population randomization is the subject of survey sampling, as treatment randomization is of experimental design. Here we note the common roots of both (1.7, 1.8, 2.7).

From accepting distinct values of e_i for the elements the need for randomization of treatments follows as strategy against the biases in treatment means that would arise from the selection of subjects into treatment groups. Then randomization over the population is needed as strategy against biases in the selection of subjects, since we cannot assume that the their e_i values have been randomized over the population.

Why randomize? This question still provokes many theoretical discussions and differences that are much too profound for us here. But there is a great deal more agreement in the practice of good statisticians. Thus I shall risk a summary that is both simple and sound. (1) Probability and statistical theory are based on random variables. (2) Populations, physical universes do not come to the researcher prerandomized. Gross irregularities and clustering characterize the populations of social research. (I believe this is generally true throughout nature.) (3) Therefore researchers must accomplish what nature did not: Randomize their samples. (4) This can only be done with mechanical randomization, such as tables of random numbers [Kish 1965, 1.7]. (5) Thus point 4 follows because the haphazard choices of human selectors are not random, even when they are honestly trying.

This last point (5) brings me back to point (1), because random variation is also needed for the Laws of Large Numbers, so that statistics $\bar{y}$ approach the parameters $\bar{Y}$ asymptotically. Trials have shown that human judgment produces biased results even for simple tasks, tasks much simpler than selecting humans for random behavior and attitudes. Thus for selecting stones from a pile, a relative bias of $(\bar{y} - \bar{Y})/\bar{Y} = (2.34 - 1.91)/1.91 = 0.22$ was found (in ounces). Similar biases were found for selecting, for example, heights of wheat shoots. Furthermore, the biases are not reliable: They vary between observers, between dates of similar experiments, and between situations. Judgment selections tend toward the middle, avoiding extremes; therefore means of judgment samples have smaller "mean square errors" (7.1E) than random samples for very small samples (5 or 10?) [Yates 1981, 2.4–2.5; Yates 1935; Jessen 1978, 1.6; Sukhatme 1947; Sukhatme 1954, 1.8]. Furthermore, sophisticated, "statistical" balancing methods are also biased, as was shown by the early classic of modern sampling [Neyman 1934]. The need for randomization of treatments, even for "double-blind clinical trials," is discussed later (3.5).

Finally, we must admit two basic justifications for the absence of broad representation for experiments. First, it is extremely difficult to achieve; the polio vaccine experiment over the United States was a rare but interesting exception [Meier 1972]. Second, many experiments yield useful results even from small isolated groups. The simple classic model $Y_t = \bar{Y} + T_t + e$, though philosophically naive, often yields reasonable results, especially with replications (2.7, 3.1, 7.6).

1.5 STATISTICAL TESTS

The term *statistical tests* refers here briefly and collectively to statistical measures of random variability, whether tests of significance, confidence intervals, fiducial or credible intervals, or whatever. The function of statistical tests is to distinguish the explanatory effects of class E variables from the random effects of class R variables, for which the tests are designed to make allowances at specified levels of probability. In "ideal" experiments this separation of E from R is accomplished through randomization of all extraneous variables in class R, except for those controlled in class C, hence with no disturbing variables left in class D.

However, in nonexperimental research the explanatory variables of class E are confounded with disturbing variables of class D. Therefore in these cases statistical tests actually contrast the effects of the random errors of class R against the explanatory variables of class E confounded (mixed) with unknown effects of class D variables. For this reason the control and segregation of disturbing variables into class C becomes doubly important in nonexperimental research. Controls decrease random errors here also, but more important, they decrease the possible biasing effects of disturbing variables on the explanatory variables. Note the weak verb *decrease*, because the stronger *eliminate* (often used) would represent our aims but not what can be typically achieved in practice. Further, the adjective *possible* also refers to our uncertain knowledge about the actual sources and sizes of biasing effects (Ch. 3 and 4).

Suppose, for example, that in a study of schools a criterion (predictand) variable y appears to be related to treatment (predictor) variable x. The criterion may define test abilities or even much later occupational success; or it may define merely student attitudes and satisfactions. The treatments may measure the nature of instruction or of class organization in the schools. Suppose for simplicity that when schools are sorted into types A and B according to the treatment x, type A schools have a higher average of the criterion of success y than type B schools. The observed difference in success may be denoted as $(\bar{y}_a - \bar{y}_b)$. But there also exists variation in success between schools within each type, and a statistical test is conducted to measure the effect of that variability; this is often measured with the standard error of the difference, denoted by ste$(\bar{y}_a - \bar{y}_b)$. A statistical test can show whether the difference $(\bar{y}_a - \bar{y}_b)$ between the two types of schools should be reasonably considered as a chance occurrence when compared with the measure of random variability denoted by ste$(\bar{y}_a - \bar{y}_b)$. If the difference $(\bar{y}_a - \bar{y}_b)$ appears relatively large (i.e., "statistically significant"), we may conclude that the difference found in the research is not likely to be due merely to the random errors of class R variables.

When the assignment of schools to treatment types A and B is randomized, a "significantly" large difference $(\bar{y}_a - \bar{y}_b)$ can be reasonably attributed to the effects of the two types, A and B, of randomized treatments. However, in nonexperimental research, in surveys and investigations, the treatment types are not assigned at random. Hence the difference $(\bar{y}_a - \bar{y}_b)$, when found to be beyond random variability, cannot be clearly and directly attributed to the difference between the two types of treatments defined by predictor variables. Disturbing variables of class D may also be confounded (mixed) with the defined treatments. For example, the salaries of teachers and the socioeconomic level of students may also differ between the two types; these differences may account for some or all of the difference $(\bar{y}_a - \bar{y}_b)$.

The researcher may try to remove the effects of disturbing variables with various methods of control (Ch. 4). These attempts at controlling for class D variables can be followed with further statistical tests. The separation of class E from class D variables should be determined in accord with the nature of the hypotheses with which the researcher is concerned. But that separation is beyond the functions and capacities of statistical tests of "significance." Their function is not explanation; they cannot point to causation. Their function is to ask, "Is there anything in the data that *needs* explaining?" Is the difference (or relationship) great enough to place confidence in the result? Or contrarily, may the latter be merely due to chance fluctuations in the specific samples on which the test was made? And they must answer such questions with designated probability.

It is incorrect to allege that "tests of statistical significance are inapplicable to nonexperimental research." This allegation was clearly expressed by Selvin [1957]:

> The basic difficulty in design is that sociologists are unable to randomize their uncontrolled variables, so that the differences between "experimental" and "control" groups (or their analogs in nonexperimental situations) are a mixture of the effects of the variables being studied and the uncontrolled variables or correlated biases. Since there is no way of knowing, in general, the sizes of these correlated biases and their directions, there is no point asking for the probability that the observed differences could have been produced by random

> errors. The place for significance tests is after all relevant correlated biases have been controlled.... In design and in interpretation, in principle and in practice, tests of statistical significance are inapplicable in nonexperimental research.

In a criticism, McGinnis [1958] shows that the separation of explanatory from extraneous variables depends on the type of hypothesis at which the research is aimed. See also Kish [1959]:

> The control of all relevant variables is a goal seldom even approached in practice. To postpone to that distant goal all statistical tests illustrates that often the perfect is the enemy of the good.... In this sense, not only tests of significance, but any comparisons, any scientific inquiry other than the ideal experiment would be inapplicable. Such defeatism is indeed advocated by enemies of the social sciences.

Indeed it would apply to all nonexperimental research. It is true and inconvenient that class E explanatory variables are confounded with class D disturbing variables. Nevertheless, statistical tests can be used to separate the effects of class R random variables from the other variables.

Statistical tests of significance have definite functions, but those functions are limited. Their limitations should be emphasized along with their common misuse for measuring the strengths of relationships between explanatory variables. The results of tests of significance are functions not only of the magnitudes of relations but also of the numbers of sampling units used and of the efficiency of design. In small samples meaningful results may fail to appear "statistically significant," whereas in large samples the most insignificant relationships can appear "statistically significant."

The word *significance* conveys a sense of importance, of meaning, in common parlance; its use in *statistical significance* amounts to a statistical pun whose effects are confusing. Tests of significance are particularly ineffective as they are commonly used in social research: to test null hypotheses of zero differences, or null relationships. Such hypotheses are trivial reflections of the actual aims of social research. Independence between social variables hardly ever exists, and it is seldom even approached. If we have a large sample or a complete census on our tapes, we can almost always find the relationship between any two variables to be greater than zero. This is a consequence of the highly multivariate and complex interrelationships of social variables [Kish 1959].

However, statistical tests have important functions in assessing the results of empirical social research. Specifically, the standard errors ste$(\bar{y}_a - \bar{y}_b)$ of differences $(\bar{y}_a - \bar{y}_b)$ found for pairs of treatments A and B can be used to assess the probability intervals of random effects *R*. With these intervals we aim to separate random effects *R* from the effects of explanatory variables *E*. Both the valid and the efficient management of those standard errors and intervals are the two chief purposes of statistical design.

1.6 AN ORDERED LIST OF RESEARCH DESIGNS

So far we have posed the choice between the classic "ideal" experiment and the classic survey sample as a basic strategy for research designs. These two designs occupy almost entirely the literature of statistical designs. But between these two pure types we find in actual practice frequent use of compromises with a great variety of other designs. I have forced that diversity into 10 types of designs and into an ordered list. I hope to delineate with this scheme the differences between experimental (analytical) studies and enumerative (descriptive) studies; and to link the experimental/survey dichotomy of this chapter to the greater diversity found in actual practice and in the following chapters. We begin with the extreme type of enumerative survey (S1), where representation is most feasible and desired, and end with the "ideal" experiment (E10), where randomized treatments are most feasible and needed. Observational studies are placed near the middle (05, 06). The 10 types of design listed in Table 1.6.1 are more thoroughly discussed in Chapters 2–4.

TABLE 1.6.1. An Ordered List of Research Designs

S. Survey samples: representation with probability selection
   1. Means and totals for frame populations
   2. Means for domains; inferences to other populations
   3. Comparisons of domain means; analytical uses; controls
   4. Multivariate analysis; regressions; categorical analysis

O. Observational studies: realism of treatments and effects
   5. Replication over several sites of treatment/control
   6. Depth study on single site of treatment/control

E. Experiments: randomized treatments, double-blind trials
   7. Internal replication over several sites
   8. Replication or combination of single experiments
   9. "Fixed" and "mixed" models
   10. Classic random models of controlled causal systems on single sites

We begin with classic survey sampling for estimating totals and means for the entire population contained within the selection frame (S1); this occupies most of the literature of survey sampling. However, most surveys must have purposes and uses that transcend those narrow aims. Means of subclasses in the sample are commonly used for estimating means for domains within the population (S2). This is not unrelated to the need for and practice of using samples for inferences beyond the frame population to the target population and further (1.8, 2.3). Often we also find analytical comparisons of domain means (S3); these may be enhanced with controls for disturbing variables and with statistical techniques, such as analysis of variance and covariance, borrowed from experimental designs. Further, the full use of multivariate statistics of all types for causal analysis of survey data (S4) has spread along with computing hardware and with computing and statistical literacy (7.1). Randomization of treatments and fully specified models may not be feasible; but they may be approached with more realistic and flexible theory and philosophy (2.7, 7.6).

At the other extreme, the classic random model (E10), fully specified and controlled, can seldom be brought or enticed into the laboratory. This is especially true in the social sciences, and "double-blind clinical trials" serve as reminders of some of the difficulties with all experiments on humans. The "fixed" or "mixed" models (E9) versus random models in the literature of experiments refer to representing finite populations. However, we need broader bases because experiments often leave wide gaps of inference and because they are conducted on narrowly and/or poorly defined populations. Models can seldom be specified so well that a stable causal system can be reasonably assured. Experiments need replications over subpopulations (E8) of great diversity to expose their results to possible "falsifications," so that the boundaries of the inferential span of a theory can be mapped with greater confidence (7.6). Numerous and prompt replications serve as safeguards, however imperfect, of the narrow bases so common in medical experiments. On the contrary, in the social sciences, replications are much too rare and imperfect; and many results remain in splendid isolation, neither refuted not confirmed. To facilitate and speed broader inference we should advocate *internal replication* (E7) whenever feasible: replication of the experiment itself over several sites that invite (maximize) the possibility of "falsification." If the experimental results are consistent over all the conflicting sites, they thereby acquire broader inferential boundaries (Section 3.1).

Observational studies are placed between surveys and experiments, and they have some resemblance to each, as well as differences with both. Their great variety defies clear definitions, but we may agree that though they lack the randomization of treatments in experiments, some control of treatments may be aimed for, and that this may be more feasible on restricted sites (O6). On the other hand, with replications over several sites (O5), we can better spread the results over the population. But we need not randomize over the entire population; we may instead concentrate the study in putative critical subpopulations.

The needs for descriptive, enumerative surveys versus analytical studies of causal systems are commonly counterposed in the literature. Sometimes actual situations do arise where one need clearly predominates, and these can and should be recognized with some care and judgment, and the proper design used. However, in most situations the two kinds of needs, enumerative and analytical, are not immediately and clearly distinct. For most research projects all the 10 types (and their various modifications) should be examined and discussed to see which of them, or what combination of them, seem most feasible and desirable, least expensive, and most promising of rich inferences.

1.7 REPRESENTATION AND PROBABILITY SAMPLING

In a series of four articles on "Representative Sampling," Kruskal and Mosteller [1979–1980] list six categories of its meanings that confront us in the nonstatistical literature: (1) general, unjustified acclaim, approbation for the data; (2) absence of selective forces; (3) mirror or miniature of the population: the sample has the same distributions as the population; (4) typical or ideal case; (5) coverage of the population: samples designed to reflect variation, especially among strata; (6) probability sampling: a formal sampling scheme to give every population element a known positive probability of selection.

We have been faced by all this variety of meanings in books and articles, scientific and nonscientific; in newspapers; in the courts; etc. We should not be surprised by this ambiguity and variety, because such is the fate of other big concepts like statistics, species, mathematics, physics, etc. But we should not accept all the proposed meanings as equally appropriate, and neither do the two authors.

The principal purpose of representation is to allow us to make inferences (rationally and probabilistically, though not with certainty) from samples to target populations; then to infer even higher to broader populations of inference (2.1) "General acclaim, approbation" (1) and "absence of selective forces" (2) are only part of the atmosphere, mere accompaniments of our purpose, and cannot serve as definitions of representative sampling. The "typical or ideal case" (4) as representative is treated separately later (3.1A).

The *aim* of representative sampling is to make the sample a miniature so as to mirror and to represent the population with *similar* distributions; such representation serves the purpose of our intended inferences. The plural in "distributions" refers to representing not only one statistic (like the sample mean), and not only a single variable, but the multivariate distributions of all variables and their functions. A rigorous definition poses difficult challenges for mathematical statisticians. Thus I accept meaning 3 as only stating the *aims* of representative sampling, and meaning 5 as a less clear statement of it. But it does not describe a method for reaching those aims.

> *Representative sampling* is a term easier to avoid because it is disappearing from the technical vocabulary. At different times it has been used for random sampling, proportionate sampling, quota sampling, and purposive sampling. In general, it often denotes *the aims* of representing a population well with a sample; and this is the sense of the terms *populating sampling* and *survey sampling* in our vocabulary. [Kish 1965, 1.6]

Probability sampling (6) denotes the only feasible *method* recognized by survey samplers in most practical situations to achieve the aim stated in meaning 3. Thus we may clarify the connection between probability sampling with randomization as the practical, empirical, objective method (means) for achieving the stated aims (ends) of representative sampling. Probability sampling requires known positive (nonzero) probabilities of selection assigned to each element, assured with operational, mechanical randomization over the population in the selection frame [Kish 1965a, 1.7]. We also know that we must accept many empirical data obtained without representative sampling and without probability sampling. But that is no justification for not keeping the definitions straight.

Please note that for aims I used *similar distributions* rather than the same distribution in order to admit sampling variation. This brings us to the problems of *measurability* of sampling variation, and of *large samples* to obtain sampling *consistency*. These are valuable additional criteria for "good" probability samples [Kish 1965a, 1.6]. Consistency refers to the approach of statistics (sample values like $\bar{y}$) from large samples to the corresponding population parameters ($\bar{Y}$), i.e., the absence of large statistical bias. Measurability refers to the operational capability for computing, from the sample data themselves, sampling errors for the statistics ($\bar{y}$) around the $\bar{Y}$ (7.1).

1.8 MODEL-DEPENDENT INFERENCE

Justifications that would make probability selections unnecessary are presented in this section, together with my skeptical personal views about those

TABLE 1.8.1. Some Names Used to Contrast Two Sampling Approaches

Model-dependent	Population-bound
Model-based	Design-based
Modeling	Survey sampling
Population free	Representation
Theoretical	Randomization
Mathematical	Physical, empirical
Model-dependent	Model-based

All the terms in Table 1.8.1, and more, have been used for these two approaches, and not only in these seven pairs. Theory, mathematics, and models must be used in any approach, and that is why Hansen, Madow, and Tepping [1983] prefer the last pair, though *model-based* is commonly used for the other approach. The first pair seems to me the easiest to remember for contrasting the two approaches and with the least prejudice.

justifications. Most of my skepticism would be shared by most survey samplers, I believe, but not by all statisticians. My views also include recognition of the frequent need for accepting data, design, and inference without the benefits of probability sampling (1.3). I note also "four obstacles to" (2.7) and "substitutes for" (3.1) probability sampling.

1. *Uniform Models.* Assumptions of a uniform model are common and pervasive, and they have widespread and strong roots in statistical analysis. These assumptions may have diverse bases, ranging from simple and naive ignorance of the problem all the way to sophisticated models of "superpopulations" to which the sample may be related, traced, or attributed. These models with "exchangeability" would make randomized selections unnecessary. They may be called "model-dependent" theories as opposed to a "population-bound" approach of survey samplers like me (see Table 1.8.1). Cassel, Sarndal, and Wretman [1977] and Ericson [1969] may be good sources for the former, and Kalton [1981] and Hansen, Madow, and Tepping [1983], for the latter.

Model-dependent sampling theories have great mathematical appeal and close links with classical statistical theories of random variables; most notably and tersely in mathematical statistics, where I.I.D. means "n random variables independently and identically distributed" (1.4). The difficulties arise when we leave these comfortably randomized and specified models and superpopulations for the unknown, irregularly clustered populations of the real, empirical world. If these would obey the assumptions of the theoretical models, then the arduous and costly efforts of randomized selections and probability designs would not be necessary. On the other hand, probability designs alone without models are not sufficient to reach beyond the frame and beyond the target population all the way to populations of inference (2.1). Also they are often not at all feasible (1.3).

2. ***Well-Specified Models.*** A sophisticated, theoretical argument from the model-dependent side appears in econometric terms: failures of models are caused by "misspecification," or because the model was "incompletely identified." This implies that a population-free basis could be attained by a better-specified or a completely identified model. However, in the real world that ideal state simply cannot be attained. For example, take the simple model $d = 0.5gt^2$ for the distance covered by a freely falling body. The constants 0.5 and 2 are yielded simply and precisely by the model. But the exact value of g varies in an irregular manner that is *not* identifiable by any model: It must be surveyed carefully over the earth's surface with hard empirical work requiring great precision. Research results in the social sciences differ from this simple model only because they are subject to larger variations and, especially, in many variables.

More specification is commonly introduced by statisticians into sampling designs with "stratification" and with "blocking" into experimental designs. These methods control some of the disturbing variation that is assignable to stratifying (blocking) variables. But all empirical evidence shows clearly and repeatedly that these methods yield only partial controls, and usually only modest reductions of errors; most or much of the survey (experimental) variation remains within the strata (blocks). Relations of the best available stratifying (blocking) variables with the study variables leaves much of the variation unaccounted for, unspecified, and unidentified.

3. ***General Attacks on Statistics.*** A similar attack on probability sampling blends into attacks on the utility of statistics in general: "If the descriptions and inferences from incompletely specified models only give averages for populations, what good are they for inferences to and predictions for individuals or for homogeneous, identifiable groups?" It is true that statistics deal with averages for aggregates and yield only probabilities, often only vague results, for individuals. Samples do yield inferences about subpopulations and domains (2.3, 7.3); but practical limits (in numbers of cases and units) are reached at levels of aggregates much above the level of individuals, who typically cannot be predicted with precision. (And some may add, "Amen, and so be it.")

"If I need statistics," said some physicist, "I just take more observations instead." More observations reduce sampling errors, and they imply spending one's way out of the uncertainties (both probablistic and theoretical) of statistics. Others say, "If I need statistics, I know I've done a bad experiment." But this implies the reverse of the other: that substantive problems and measurement biases swamp statistical errors. These attacks are common in the social sciences, which endure attacks against statistics in favor of clinical methods, depth, and participant observations, as well as intuition and introspection (2.7). But for the substantive problems neither more nor better statistics are substitutes; we can admit that without going into all those problems that are not central to a book on statistical design.

CHAPTER 2

Analytical Uses of Sample Surveys

It is characteristic of those matters in which something is known with exceptional accuracy that, in them every observer admits that he is likely to be wrong, and know about how much wrong he is likely to be.... It is an odd fact that subjective certainty is inversely proportional to objective certainty. Bertrand Russell, The Scientific Outlook.

Scientists know nothing for certain. Gerald Piel, Science, *17 Jan. 1986.*

2.1 POPULATIONS OF ELEMENTS AND SAMPLING UNITS

Textbooks on sample surveys present adequate treatments for a limited set of statistics from clearly defined populations, and we shall not repeat them here. However, those treatments are deficient for our purposes in two respects. First, concentrating on statistics of simple means and aggregates, they neglect comparisons and other analytical statistics with which we deal in this chapter and in Section 7.1. These brief pages must suffer between the double pressures and handicaps of being inadequate for covering their vast subject and of being too terse, dense and curt. I fear that this chapter (unlike the others) is not "self-contained", and is difficult for readers without a course in survey sampling. But references guide the readers toward an ample coverage of this vast subject [Yates 1981; Cochran 1977; Kish 1965a; Hansen, Hurwitz, and Madow 1953].

Second, sampling theory assumes populations that are more simply and clearly defined than those we need for subjects of representation and inference. Let us agree that for most surveys it is difficult or impossible to make the samples entirely representative of the desired populations. Beyond sampling variations are the diverse divergences that may bias the selection, such as defective frames and nonresponses. Distinctions are usually drawn between the frame population and the target population, but I find it useful to distinguish four populations to account for the diverse types of discrepancies between achieved samples and the ultimate aims of inferences based on their results: the survey, frame, target, and inferential populations (Figure 2.1.1).

These four kinds of populations serve to distinguish three types of discrepancies common to most surveys. One or another of these types may be negligible on some surveys and then the lucky researcher may collapse two populations into one. Statistical inference from probability samples to survey populations uses sampling errors, which can be measured probabilistically from the data yielded by the samples themselves. This important property of sampling errors is called "measurability" (1.7, 7.1). Thus it differs from the three other kinds of inferences, which require, alas, models and/or data from outside the sample.

The survey population differs from the population of the "frame" from which the sample was selected, because of losses due to total nonresponses (refusals, not-at-homes, etc.) and due to "item nonresponses" (i.e., items missing from accepted interviews, denoted as "not ascertained") [Kish 1965a, 13.4]. But the frame population can also differ from the intended "target population," because of deficiencies in the frame. These deficiencies are called "noncoverage," or missing units or incomplete frame, and may be the net result of undercoverage minus overcoverage. Figure 2.1.1 illustrates those deficiencies, nonobservations, but the magnitudes of these can vary widely. For example, a survey in a developed country may have small noncoverage but very large nonresponses, both not-at-homes and refusals, especially for incomes; on the other hand, fertility surveys in developing countries may have small nonresponses from the selected women but may suffer from large noncoverage of dwellings. Estimating the magnitude on noncoverage is difficult, because it requires going beyond the sample and the frame to outside data, like a census, register, or quality check [Kish 1965a, 13.3]. But for nonresponses the magnitudes can be estimated with careful checks of the selected sample, though not their effects. Different from unplanned noncoverage are the deliberate "exclusions" for practical and economic reasons of part of the population.

However, sample results are also used beyond the originally designated target population for inferences to a wide variety of other populations, which differ from the target in kind, scope, location, time, etc.; these are necessary and welcome uses of research data. For example, a sample of Michigan 1986 may be used for inferences to Ohio or to the United States in 1986 or in 1990.

Inferences from the survey to the frame population, and even to the target population, have bases in data. At least they can and should have such bases, though these may need preparation and effort. The size of nonresponse can be measured and the size of noncoverage estimated. Studies can be made to

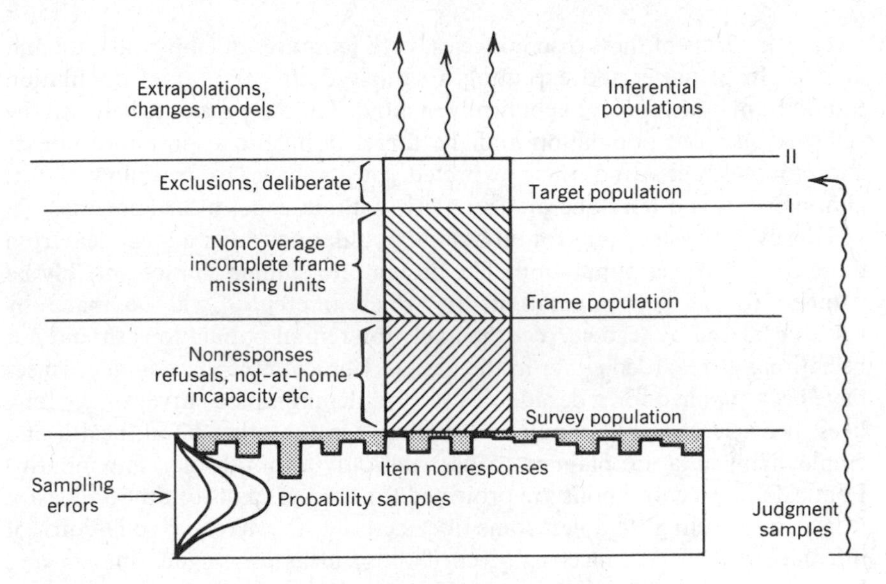

Figure 2.1.1. Discrepancies between four populations.

Probability samples underlie the achieved survey population, but two discrepancies come even between them: sampling errors and item nonresponses. Both of these differ greatly among variables and the amount of item nonresponse is shown as differing greatly among variables. For both of these discrepancies the sample responses serve as bases; sampling errors are computed from them; and they are used for "imputing" or weighting for item nonresponses.

Thus probability samples are shown as a broad and solid foundation for the survey population, on which to build the structure of the inference above it. For the discrepancies beyond the survey population one must go beyond the sample data, with the help of implicit or explicit data. The span to the frame population is due to *total nonresponses* of diverse kinds (refusals, not-at-homes, etc.); the size of nonresponses may be estimated from sample records (with effort and care), but estimating their effects needs models and auxiliary data. The size of *noncoverage* can only be estimated with models or from checks with outside sources, yet this portion also belongs to the target population. This may also include a defined and deliberate *exclusion* from the coverage.

Furthermore, sample data are also used for inferences beyond the target populations, and these are many, various, and ill defined. "Superpopulations" of sampling theory are not only among these, but behind all these inferential populations. These model-dependent inferences (1.8) are too often merely implicit. Even more vagueness describes the path of judgment samples directly to the target population, and such vagueness is indicated by the thin, wavy, population line, as for the extrapolations to vague inferential populations.

reveal the effects of these discrepancies, with joint uses of empirical data and models. In planning and executing a sample design the target population should be planned for and kept firmly in mind. The discrepancies between the achieved sampled population and the target population—in nonresponses and noncoverage—and their expected effects have to be allowed for; although this is often done only vaguely, without exact measurements.

However, the situations for inferential populations differ a great deal from those for the target population: The design of a sample cannot possibly be planned for all the populations to which inferences will be made by researchers and by readers. Yet the likely inferential populations should not be entirely disregarded either in the design. For example, decennial censuses have been planned for a decade of use; and demographic surveys have long lives, because of stable population models. On the other hand, health and employment data are planned to describe only a month or a few months. Political and electoral polls are projected for limited spans of time and space (6.1). In planning samples some ideas about the likely populations of inference should be conceived, even if those ideas are vague, limited, and incomplete. This should be true of all research endeavors.

If the inferences from survey populations to diverse inferential populations require models and judgment, why not use models for inference from any judgment sample directly to the target population? These questions are raised both by naive practitioners and by sophisticated theorists, though in different idioms. The differences between probability and judgment samples are basically similar to those between the difficult and uncertain steps of the empirical sciences and the imaginative flights of pure speculation. These differences are too vast for discussions here, but we refer to those of Sections 1.4, 1.7 and 1.8. Inferences from a ill-defined or undefined "sampled" population to a target population and to higher inferential populations would require flights of imagination, or stronger models than usually can be justified in social research. On the other hand, we must also admit that probability samples for desired target populations are often not feasible, and to be realistic and honest, we shall look at some substitutes in Section 3.1.

The *elements* of a population are the elementary units for which information is sought, and they comprise the population to which inferences are to be made. Elements are the units of analysis determined by the research objectives, and the population is defined jointly with the elements.

Sampling units contain the elements and they are used for selecting elements into the sample. In *element sampling* each sampling unit contains no more than one element. But in *cluster sampling* the sampling units are *clusters* of several (or many) elements (e.g., persons in households, households in blocks or counties, students in classes). For example, classrooms, schools, or school districts may all be clusters with each used at three successive *stages* for selecting a sample of students. In *multistage sampling* a hierarchy of sampling units is used for selecting the sample, so that each element must be identified with (belong to) a single sampling unit at each stage. Thus the population also becomes an aggregate of the sampling units at each stage, and the sampling units at each stage comprise complete partitions of the population at that stage.

Listing units (listings) are used to identify and select sampling units from *lists*. Instead of actual lists we may need to use a *frame* as a procedure for selecting units. *Frame problems* (deficiencies) denote the lack of a one-to-one correspondence (L–S) from listings (L) in the frame population to the sampling units (S) in the target population. Such problems can be sorted into four classes (with 0 for absence): (1) missing units (0–S), (2) blanks (L–0), (3) replicate listings (L–S–L), and (4) clustered units (S–L–S) [Kish 1965, 2.7, 11.1–11.5].

Observational units are sources of information, and they are called respondents in interview surveys. They may be distinct from elements; e.g., for a survey of children the observational unit may be mothers (for births and for health data) or teachers (for education).

One survey may yield statistics about several diverse populations. (1) *Different contents*, e.g., surveys of crime or of home accidents can yield data about separate incidents, persons harmed, or families, households, or homes with incidents. (2) *Different units*, e.g., consumer data may be tabulated by persons, spending units, families, or households. (3) *Different extents*, as in the analysis of subclasses (2.3). (4) *Different periods* may be covered by the same survey (6.1).

2.2 INFERENCES FROM COMPLEX SAMPLES

Standard methods of statistical analysis have been developed on assumptions of simple random sampling (srs). Assuming independence for individual elements (or observations) greatly facilitates the mathematics used for distribution theories of formulas for complex statistics. "Given n random variables" is either stated explicitly or assumed implicitly, for most measures of reliability, such as $\sigma/\sqrt{n}$ for means, etc. This concentration is represented by row A in Figure 2.2.1.

However, independent selection of elements is seldom realized in practice, because much research is actually and necessarily accomplished with complex sample designs. It is economical to select clusters that are natural groupings of elements, and these tend to be somewhat homogeneous for most characteristics. The assumptions may fail mildly or badly; hence

Selection methods	Statistics: 1 Means and totals of entire samples	2 Subclass means and differences	3 Complex analytical statistics, e.g., co–efficients in regression
A. Simple random selection of elements			
B. Stratified selection of elements		Available	Conjectured
C. Complex cluster sampling		Available	Difficult: *BRR, JRR, TAYLOR*

Figure 2.2.1. The present status of sampling errors. Row 1 is the domain of standard statistical theory, and column 1 of survey sampling [Kish and Frankel 1974).

standard statistical analysis tends to result in mild or bad underestimates in the lengths of reported probability intervals. Overestimates are possible, but rare and mild. Those complexities are discussed below briefly, and in some detail in textbooks and many articles on survey sampling [Kish 1965a, 5.1, 14.1, 14.2; Kish 1957; Kish and Frankel 1974; Verma, Scott and O'Muircheartaigh 1980]. These (and many others) show how neglect of the complexities of design often lead to grossly understated sampling errors and intervals, and thus to overstated confidence in sample results.

Publications on survey sampling typically concentrate on providing estimates of means and totals for the frame population, estimates of the form $\bar{y} = y/n$ and $\hat{Y} = y/f$, where $y = \Sigma y_j$ is the sample sum for the n elements in the sample, and $f = n/N$, the uniform sampling fraction. This concentration is represented by column 1 in Fig. 2.2.1. The published methods also include estimators for variances and standard errors, $\text{ste}(\bar{y}) = \sqrt{\text{var}(\bar{y})}$; related measures of sampling errors; and methods for computing these from the sample data. The chief function of these statistics is to provide probability intervals of the type $\bar{y} \pm t_p\text{ste}(\bar{y})$. Standard errors are provided for a great variety of selection methods (clustered, multistage, stratified, etc.) and for diverse methods of estimation (ratio, regression, difference) for either the mean or the total. These include weighted means $\Sigma w_j y_j / \Sigma w_j$ to compensate for unequal probabilities of selection and for modified methods of estimation.

Thus the methods of survey sampling concentrate on simple statistics from complex selections (column 1), whereas standard statistical analysis (row A) deals with complex statistics but only from simple random selection of elements. However, we often need complex statistical analyses from data that come from complex samples. Complexity can take many forms in either dimension; hence it would be difficult to list all their possible combinations. But we can conveniently and usefully divide complex selections into two classes: stratified element sampling and clustered selections (usually also stratified), which have much more drastic results on sampling errors. Also subclass means and their differences have much simpler treatments than more complex statistics, like regressions.

For means based on the entire sample, or for similar totals, two simple generalizations are commonly known and widely useful. Proportionate stratified element sampling generally reduces variances, but usually only mildly. Cluster sampling generally increases variances either mildly or badly, depending on several factors; and those increases survive the ameliorating effects of the stratification that usually accompanies cluster sampling. Both the reductions of stratification and the increases of clustering are expressed as ratios of actual ("true") variances to the variances (σ^2/n) of simple random (independent) samples with the same number n of sample elements. These ratios are well known as "design effects" (deff or deft^2), which are justified theoretically and documented empirically in the literature of survey sampling. These generalizations assume equal probability selection methods ("epsem") and "self-weighted" statistics; unequal probabilities and corresponding weights introduce complications that obstruct such easy generalizations (7.4).

Methods of survey sampling can be applied fairly readily to subclass means and to their comparisons (Section 2.4). Stratified element sampling (B2) tends to *reduce slightly* the variances for subclass means: They are reduced compared with simple random sampling (srs), but only slightly compared with effects on the mean for the entire sample. In clustered sampling (C2) the sampling errors tend to be increased compared with srs, but effects for subclasses tend to be less than those for the entire sample (Sections 2.4, 2.6).

For complex and analytical statistics it is difficult to make sweeping generalizations; because the subject seems too complex for this early chapter, it is postponed (7.1). There are many diverse kinds of statistics and theoretical results are few, because the mathematics are difficult. Computations with modern machines and methods, though not easy, are possible. For clustered samples (C3) those methods are denoted as BRR (balanced repeated replications), JRR (jackknife repeated replications), and Taylor (or linearization) methods; and the increases of variances for clustered samples have often been found to be variable, irregular, and considerable (7.1). For stratified element sampling the effects (reductions) are conjectured to be small, even negligible.

The preceding discussions deal with sampling errors, confidence intervals, and similar probability statements. These measures of the variability of descriptive statistics may be called *inferential statistics*, or "second-order" statistics; and they depend heavily on the methods of sampling, technically on the joint probabilities of selection of elements (7.1). On the other hand, we also have a useful generalization: The *descriptive statistics*, or "first-order" statistics, are little affected by the complexity of selections. Hence the *descriptive statistics*—like means ($\bar{y}$), element variances (s^2), coefficients of regression (b_{yx}), or correlations (r_{yx}), etc.—*computed "properly" from large samples are "good" estimates of corresponding values in the population* ($\bar{Y}$, S^2, B_{yx}, R_{yx}, etc.) (7.1). By "good" we mean technically "consistent," or nearly unbiased, for "large" samples—large in numbers of sampling units. Computed "properly" means weighting inversely to probabilities of selection, if these are unequal. For the preceding sweeping generalizations there exist some theoretical grounds and more empirical evidence, but little in the way of general discussion [Kish and Frankel 1974].

2.3 DOMAINS AND SUBCLASSES: CLASSIFICATIONS

Analyses based on subdivisions of survey samples may be the most common method of results for social research today. Reasons for that popularity are discussed later and placed in contrast with some alternatives (3.1 and 4.2). There is confusion about the words *subclasses* and *domains* for subdivisions. Let us use *subclasses* for the subdivisions (partitions) of the sample and *domains* for the corresponding subdivisions (subpopulations) of the population. The subdivisions represent categories of some variable, like age, occupation, region, etc., and sometimes two (perhaps more) variables such as sex–age subclasses or region–occupation subclasses. Means of subclasses are used for estimating the means of corresponding domains; other subclass statistics estimate corresponding domain parameters in the population.

Sometimes we may be concerned with subclasses based on behaviors rather than age, sex, and other predetermined characteristics; for example, subclasses of marriage status (single, married, divorced, widowed) as bases for observing other behaviors or attitudes (health, satisfaction, etc.). The subclasses serve as denominators for means and generally as predictor (causal, stimulus) variables in statistical analyses. The designation of subclasses depends on the needs and aims of the researcher to explain differences among domains in the population; thus expressed attitudes (e.g., Republican, Democrat, independent, etc.) may also be used for subclasses. Furthermore, subclasses are used not only as predictor variables, but also as controls for disturbing variables (4.2). Note that we use the *subclasses as denominators*

Types of Classes	Sizes of Classes		
	Major	Minor	Mini
Design classes	Major regions, provinces	50 states of United States	3000 counties of United States
Mixed classes	Partial segregation: natural resources, cultural, ethnic, or mixed types: regions × age		
Crossclasses	Five-year age groups Major occupations	Single years of age Occupation × education	Years of age × education Age × education × income

Figure 2.3.1. Classification of Domains and Subclasses (with Examples)

for means denoted as $\bar{y}_c = y_c/n_c$ and for other subclass statistics. However, if we are discussing relative sizes of subclasses as numerators in the sample, such as n_c/n or y_c/y, we would call them "shares" of the entire sample.

The statistical sampling aspects of design can be facilitated with rough classifications both of types and of sizes for domains and subclasses (Figure 2.3.1). First note three rough *types of domains* (*and subclasses*).

1. *Design domains* designate subpopulations for which separate samples have been planned, designed, and selected. The combination of design classes forms the entire sample, usually as a sum of independent samples. In national samples, examples are the major regions and the urban and rural domains, and these are composed of entire strata of primary sampling units.
2. At the other extreme are *crossclasses* that cut across the sample designs, across strata, and across sampling units. These are the most commonly used kinds of domains and subclasses—e.g., age, sex, occupation, education and income classes, behavior and attitude types, etc. They were not separated into design domains because information was not available on these variables, or because they were ignored.
3. Between the two extremes, but less commonly used than the two dominant types, are *mixed classes* of diverse kinds. They were not separated by the design, but they tend to concentrate unevenly in the sampling units and in the strata. For example, occupations such as fisherman, farming specialties, miner, and lumberjack, which are segre-

gated by natural forces; or ethnic groups segregated by social forces. In both cases the segregation may be prevalent, but neither complete nor available as auxiliary data for design needs.

The *sizes of domains* also influence the choice of methods for design and estimation; hence a crossclassification of the preceding types with classes based on sizes of domains also seems useful. This classification is stated here roughly to orders of magnitudes, with descriptive names assigned for ready reference. Although the boundaries are arbitrary they are useful, because different size classes need different sampling strategies.

1. Major domains comprise perhaps 1/10 of the population or more. Examples are major regions for design domains and 10-year age groups or major occupational categories for crossclasses. For major domains reasonable estimates can be produced with standard methods from probability samples.
2. Minor domains comprise perhaps from 1/10 to 1/100 of the population. Examples are populations of the 50 states of the United States; or single years of age; or two-fold classifications of major domains like occupation by education; or regions (designed) by education (crossclass).
3. Mini-domains comprise perhaps from 1/100 to 1/1000 or even to 1/10,000 of the population. Examples are populations of the over 3000 counties of the United States or a three-fold classification of age by occupation and by education. For mini-domains usually (and often for minor domains) the sample bases are too small for reliability, hence standard methods of estimation are inadequate, and new methods are needed (5.3) [Purcell and Kish 1980].
4. Rare types, comprising less than 1/10,000 in the population, are problems for which samples of an entire population are useless, and separate lists and methods are needed [Kish 1965a, 11.4].

2.4 OVERVIEW OF SUBCLASS EFFECTS

Survey sampling methods deal principally with entire samples selected to represent some specified populations, and discussions of subclasses in textbooks are too limited. Therefore let us have here a brief overview of diverse effects that arise when subclasses of the sample are used for inferences about corresponding domains of the population. We are concerned especially with *major crossclasses* because they are the most used and most useful and because the sweeping generalizations below apply most readily to them.

Selecting subclass members from a sample has the effect of assigning zero values to all variables of nonmembers. Hence the effects are similar to nonmembers appearing as blanks in the selection frame; the proportion of blanks increases as subclass proportions decrease.

1. Selection probabilities are preserved for individual elements in subclasses. The probabilities P_i assigned to member elements are unaffected by the zero values assigned to all variables of nonmembers when creating subclasses. Assigned weights proportional to $1/P_i$ can also be used for the subclasses.

2. Sample sizes n_c become highly variable for crossclasses. Zero values for all nonmembers have the same effect as blanks in the selection process. When crossclasses become smaller, the variability increases greatly and size controls designed for the entire sample tend to become lost for small crossclasses.

3. Estimates of means and totals retain their forms for subclasses. The unbiased nature of simple totals $\Sigma y_j/P_j$ is retained with the undisturbed selection probability (1). The ratio means $\bar{y}_c = y_c/n_c$ retain their sturdy consistency until the variability of size (2) in the denominator becomes too high for small crossclasses.

4. Variances of means and totals become greatly affected as the sizes of subclasses are decreased. The main effects are increases of variances in rough proportion to decreases in subclass sizes. But these simple srs effects are modified by design effects in complex samples. We shall discuss later the different effects in stratified element sampling (2.5) and in cluster sampling (2.6); and the effects are different for crossclasses and for design classes. Very briefly, for means of design classes the average effects tend to resemble the effects on the entire means—usually modest reductions of variances for proportionate stratified random element sampling (pres); but for cluster sampling, increases of variances may be small or large. However, for crossclasses both the reductions of pres and the increases of cluster sampling tend to disappear from the variances; and for crossclasses representing small proportions the variances approach those of srs.

Most important, but also most difficult to quantify, are the *likely effects on the ratios of biases to variances in the mean square errors*. This change of the ratio of bias/σ in domains in neglected often but not entirely, and the basic argument (model) is simple [Kish 1965a, Fig. 13.1.II, 13.2C]. For domains the standard errors increase by the ratio $\sqrt{n_c/n_t}$ as the sample size decreases from n_t for the overall means to n_c for a subclass, except that for cross classes the increases are also affected (dampened) by changes in the design effects. For example, a crossclass of 10 percent of the sample, may have ste $(\bar{y}_c) =$ 3 ste$(\bar{y}_t)$, since $\sqrt{10 \times (9/10)} = 3$, if an assumed deft^2 = 10/9 almost vanishes for the crossclass. The bias for the overall mean B_t will tend to remain the same on the average for the domains. Then with $B_c = B_t$ the bias

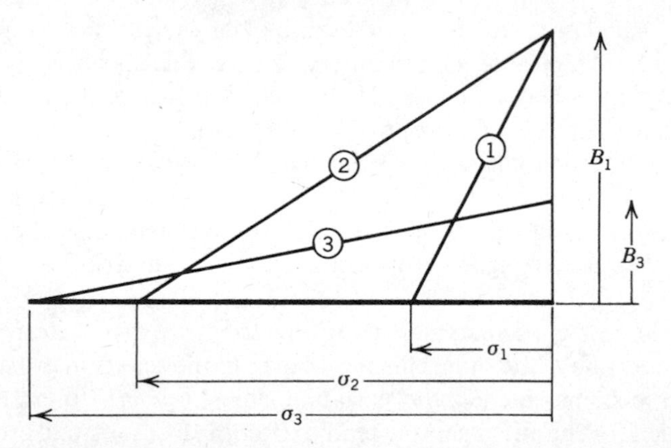

Figure 2.4.1. Variable errors (σ) and biases (B) in root mean square errors (RMSE).

The bases represent sampling errors and other variable errors (σ). For example, σ_1 may be the ste($\bar{y}_t$) for the mean $\bar{y}$ of the entire sample and σ_2 may be a larger ste($\bar{y}_c$) for a subclass mean, and σ_3 may be the ste($\bar{y}_c - \bar{y}_b$) for the difference between two subclass means.

The heights represent biases (B) and the hypotenuse denotes the RSME $= \sqrt{(\sigma^2 + B^2)}$; (see 7.2F). (1) For the entire sample the bias B_1 may be large compared with the variable error σ_1, thus taking larger samples would not decrease the RMSE_1 by much. (2) However, with the same bias B_1, but with a smaller sample in the subclass, the ratio changes and the σ_2 dominates the RMSE_2; and this is not much larger than for (1) despite a much smaller sample. (3) Furthermore, for the difference of means, the net bias B_3 may be much smaller; so that even with a larger σ_3, the RMSE_3 for the difference is but little greater than RMSE_2. This drastic change in the bias ratio B/σ tends to appear not only for differences between subclasses within the same sample, but also for differences between repeated surveys.

ratio $B_c/\text{ste}(\bar{y}_c) = B_t/3\ \text{ste}(\bar{y}_y)$. Thus the *relative maqnitude of the constant bias is diminished in the root mean square error for domains* (Figure 2.4.1).

5. *Comparisons (differences) of subclass means represent common uses of survey data.* For these differences $\bar{y}_c - \bar{y}_b = y_c/x_c - y_b/x_b$ the *denominators* x_c, x_b have several common characteristics. They generally denote categories of the same variable, and usually nonoverlapping partitions; thus they represent count (0,1) variables that distinguish members from nonmembers of subclasses. They may be self-weighting counts of n's or weighted counts; they may also represent noncategorical variables—e.g., live births y_c, y_b per total births x_c, x_b of women in age classes c and b. For k partitions of the sample each pair represents only one of $k(k-1)/2$ possible pairs.

For differences of subclass means we must again distinguish between design classes and crossclasses. For design classes, the variance of the difference is simply the sum of the variances for the means (because these are independent). *But for differences of crossclass means the variances are brought closer to srs variances. For cluster sampling the variances are further reduced* (2.6). *For proportionate element sampling the variances become practically equivalent to srs variances* (2.5). For both designs these are the typical effects on "design effects" of the covariances between crossclass means.

The *decrease in the ratio* of bias to variance we noted for subclasses is likely to be *further accentuated for comparisons* of subclasses: for bias $(\bar{y}_c - \bar{y}_b)/\text{ste}(\bar{y}_c - \bar{y}_b)$. We expect a further increase of $\sqrt{2}$ in the $\text{ste}(\bar{y}_c - \bar{y}_b)$, except for a "dampening" in design effects. But then *considerable decreases in the net bias* of $(\bar{y}_c - \bar{y}_b)$ may occur to the extent that the pairs of biases are similar (or "additive"). This compromise view is more cautious then either of the commonly held extremes: that the bias is bound to disappear or that nothing happens to it (Figure 2.4.1).

6. Designs for domains induce conflicts with the entire sample and with other domains. Conflicts arise in allocating sample sizes and sampling rates to provide adequate sample bases for diverse design domains (7.3).

7. Estimates for small domains require special techniques (Section 5.3F). These can improve estimates for minor domains, and they may facilitate estimates for mini-domains that would not be feasible at all from the sample alone. They may be used even for major domains, although their estimates can also be made from the sample alone.

2.5 PROPORTIONATE STRATIFIED ELEMENT SAMPLING (PRES)

In stratified sampling the sampling units of the population are partitioned into strata in order to select independent samples from each. For pres we assume simple random selections (srs) of elements within strata. Estimates for design domains present no special problems because for each domain we use the usual formulas of stratified element sampling (7.4.9). When the domain contains a single stratum, we are back to simple random sampling. Sampling textbooks give adequate treatments for the means and variances [Kish 1965a, 3.3].

However, most subclasses needed for statistical analyses are *crossclasses*, since they cut across the strata of the selection process. The proportions $\bar{M}_{ch}$ of crossclass members in the strata (h) leave residuals $(1 - \bar{M}_{ch})$ of nonmembers, which appear as blank selections that loosen the size control of stratification. In the hth stratum, srs selection of n_h from N_h elements with rates $f_h = n_h/N_h$ yields a crossclass with sample size m_{hc}. This number is a dichotomous random variable with expectation $f_h M_{ch}$ and variance $(1 - f_h)n_h\bar{M}_{ch}(1 - \bar{M}_{ch})$; here M_{ch} is the size and $\bar{M}_{ch} = M_{ch}/N_h$ the proportion of domain members. The effects of this variation in increased variances of crossclass means, totals, and differences we note later. Detailed treatments and derivations are available elsewhere [Kish 1965a, 1969, 1980; Cochran 1977; Yates 1960].

The effects of crossclasses on variances can be viewed most simply in *proportionate* stratified random element samples (pres), when elements are selected with the same equal probability $f = n/N = f_h = n_h/N_h$ within each stratum. This results in a *self-weighting sample* in which the sample proportions are equal to the population proportions and thus $w_h = n_h/n = W_h = N_h/N$. Thus the total sample size n has been *controlled with pres* to represent the stratifying variables found in the population N. These controls have the effect of reducing the variance from S^2/n in srs to S_w^2/n, the *within strata variance in pres*. The ratio of reduction may be called the design effect of pres:deft^2(pres) $= S_w^2/S^2$. The reduction comes from eliminating the between-stratum variances, $S_b^2 = S^2 - S_w^2$: The greater the homogeneity of elements within strata, the less is their within variance S_w^2 and the greater is S_b^2. However, it is seldom that the relative reductions $S_b^2/S^2 = (S^2 - S_w^2)/S^2 = 1 - S_w^2/S^2$ are greater than 10 percent, because finding better stratifying variables for elements (individuals) is difficult and rare.

For crossclasses even those small reductions tend to be lost in proportion to the decrease in the relative size $\bar{M}_c = M_c/N$ of the crossclass; e.g., a reduction of 10 percent for pres becomes about 2 percent for crossclasses of $\bar{M}_c = 0.2$. Thus for modest pres reductions for the entire sample the variances for crossclasses tend to srs variances, and the design effects S_w^2/S^2 approach 1. The element variances are roughly $S_w^2 + (1 - \bar{M}_c)(S^2 - S_w^2) = S^2 - \bar{M}_c S_b^2$; these become S_w^2 for the entire sample ($\bar{M}_c = 1$) and approach S^2 as the crossclass proportions $\bar{M}_c$ become small.

The increases in crossclass variances are proportional to the loss of control over sample sizes m_{ch} within strata, and the proportions $(1 - \bar{M}_c)$ of nonmembers (blanks) represent those losses. The relative sizes $m_{ch}/\Sigma m_{ch}$ of the strata for crossclasses are subject to random variations around (instead of being controlled at) their population values $W_{ch} = M_{ch}/\Sigma M_{ch}$; that is why such variances tend to approach srs variances. When reliable values for W_{ch} can be found for crossclasses and used with "ratio estimates" $\Sigma_h W_{ch} \bar{y}_{ch}$ for "poststratification," the gains of stratification (reduction of variances) of pres can be nearly recaptured. However, reliable values of W_{ch} for crossclasses are rare and mistaken values can result in biased estimates.

The tendency for the gains of proportionate stratification to disappear becomes even stronger for variances of comparisons: For *proportionate stratified element sampling the variances for differences of crossclass means tend to approach closely the variances for unstratified srs*. Thus the variance for $(\bar{y}_c - \bar{y}_b)$ tends to become simply $s_c^2/n_c + s_b^2/n_b$. The covariances between means across strata typically cancel the gains of proportionate element stratification [Kish 1965a, Section 4.5].

2.6 CLUSTER SAMPLING

Clustering generally increases the variances of sample means, and these increases can be measured with the *design effects*: the ratios deft^2 = actual variance/srs variance. The srs variance for means can be computed simply as s^2/n, but for computing the actual variance the sample design must be considered. The actual variances, hence also deft^2, depend on the nature of the variables measured, on the sample design, and especially on the sample sizes $\bar{b}$ of the achieved clusters. The sample design interacts with the distributions of variables over the population to produce *ratios of homogeneity, roh*, specific for variables. These relations can be expressed roughly, but simply and effectively, for sampling errors of means $\bar{y}$ as:

$$\text{actual (var)}/(s^2/n) = \text{deft}^2 \simeq [1 + \text{roh}\,(\bar{b} - 1)] \qquad (2.6.1)$$

Here $\bar{b} = n/a$, where n is the total size of the sample and a is the number of "primary sampling units" (PSUs) in the sample. Thus a represents the number of independent selections, and this rough formulation simply accepts the methods of subsampling and stratification actually used. Simple methods, based on primary selections (or "ultimate clusters") for computing variances and deft^2, can be found in several sampling textbooks [Kish 1965a, Section 6.5; Kalton 1979].

This expression of deft^2 in terms of roh and $\bar{b}$ separates roughly but usefully four sources of variation in actual variances.

1. Dividing the actual variance by s^2/n yields a "standardized" deft^2 from which the units of measurement, the basic variation σ^2 of the population distribution of the variable y_i, and the overall sample size n have all been removed (as "nuisance parameters").
2. The values of roh vary greatly between variables in the same sample, mostly between 0.001 to 0.2 (to give a rough idea) for most empirical data; but negative values and values closer to 1 can also be found. The statistic roh, computed as $(\text{deft}^2 - 1)/(\bar{b} - 1)$, estimates the parameter ROH (the ratio of homogeneity) for a specific variable and for the specified sample design. This differs only slightly from the classical "intraclass correlation" RHO between elements in two-stage sampling [Kish 1965a, 5.6].

3. The average cluster size of $\bar{b} = n/a$ is an important factor, since for large $\bar{b}$ even small values of roh can result in large increases of deft^2. Though $\bar{b}_t$ is fixed for any total sample size, during survey analysis the cluster sizes $\bar{b}_c$ for diverse crossclasses are decreased and do vary greatly, hence also the deft_c^2. But the roh values of a variable remain relatively stable and "robust."
4. The value of roh (for any variable) depends both on the distribution of the variable over the population and on the nature of the sample design—on the kind, size, numbers, and stratification of the sampling units used in the various stages of selection. However, separating those components of the variance (and for each of the variables) is beyond the resources of most surveys.

The main strategy for computing and for using sampling errors should aim at separating the diverse effects of the greatly different values of rohs for different variables, from the distinct effects on the computed values of deft_c^2 of decreasing subclass sizes $\bar{M}_c$ on the values of average cluster sizes $\bar{b}_c$. Hence the first step should be to *compute for the total sample size the variances, var*($\bar{y}_t$), *and the design effects, deft*$^2(\bar{y}_t)$, *for all* (*or many*) *of the most important variables* (7.1). Modern computers and programs make many such computations feasible.

The variances depend on too many factors for useful generalizations. However, values of $\text{deft}^2(\bar{y}_t)$ are much more predictable and useful, because dividing by s^2/n removes the disturbing factors of units of measurement and sample size. The computed values of $\text{deft}^2(\bar{y}_t)$ should be mostly between 1 and 6 if the average cluster size $\bar{b}_t < 100$, because most values of roh are less than 0.01 or 0.05. Thus $\text{deft}^2(\bar{y}_t)$ also serves as a rough check against wild mistakes in computations. However, some values may be lower than 1 because of sampling variation, discussed briefly later; some may be much higher, because of strong segregation in clusters of some variables (like "race" in the United States).

Design domains, such as regions, contain primary selections in separated strata; hence separate computations for each design domain could simply follow standard variance formulas. However, this may not be the best strategy, for two reasons: (1) There may be too many domains for separate computations and presentations; (2) the number of primary selections may become too small and result in unstable estimates of the variances. Thus it may be better to use "pooled" estimates of the variances; and if we assume similar $\bar{b}$ and similar average design effects for a variable over all design domains, this would be (average of deft^2)s_t^2/n_c. For differences between means $(\bar{y}_c - \bar{y}_b)$ of design domains the variances consist of the sums of variances, when the samples are independent.

For crossclasses, which occur most frequently, the effects are quite different. Crossclasses tend to cut across all strata and all clusters and hence to reduce the design effects of clustering in a drastic and fairly predictable manner: by reducing the average cluster size from $\bar{b}_t$ to $\bar{b}_c = \bar{M}_c\bar{b}_t$, where $\bar{M}_c = M_c/N$ represents the proportion of crossclass members among the N population elements. To the extent that we may assume the roh_c for crossclasses to equal the roh_t for the entire sample, we have

$$\text{deft}_c^2 = [1 + \text{roh}_c(\bar{b}_c - 1)] \simeq [1 + \text{roh}_t(\bar{M}_c\bar{b}_t - 1)]. \qquad (2.6.2)$$

Thus deft_c^2 *tends to be reduced linearly toward 1 as the crossclass proportions decrease.* This model, based on assumptions that $\text{roh}_c = \text{roh}_t$ and $\bar{b}_c = \bar{M}_c\bar{b}$, held fairly well in thousands of empirical computations across many kinds of surveys. Its imperfections pale into insignificance compared with the differences of roh values between variables, which often may be hundredfold (0.001 to 0.1) on one survey. Also sampling variations in computing $\text{var}(\bar{y}_c)$ and $\text{deft}^2(\bar{y}_c)$ directly may be less accurate than using (2.6.2) for

$$\text{var}(\bar{y}_c) = [1 + \text{roh}(\bar{M}_c\bar{b} - 1)]s^2/n_c. \qquad (2.6.2a)$$

Hence roh may be pooled for a variable from computations of deft^2 for related crossclasses that may vary in size. If the roh_c are simply imputed from variances computed for the entire sample, it is best to increase it slightly to 1.2 roh_t or 1.4 roh_t. The values of these synthetic roh_c values tend to increase slightly for decreasing $\bar{M}_c$, partly because of effects of variations in cluster sizes; and those increases are greater for those subclasses, such as socioeconomic classes, that are unequally distributed crossclasses. These relations may break down altogether for small crossclasses when $\bar{b}_c$ approaches 1, but fortunately they are not needed then because deft^2 should be near 1, with negligible effects of clustering. [See Verma et al. 1980 and Kish et al. 1976 for both methods and data.]

For differences of crossclass means the variances become $\text{var}(\bar{y}_c - \bar{y}_b) = \text{var}(\bar{y}_c) + \text{var}(\bar{y}_b) - 2\,\text{cov}(\bar{y}_c, \bar{y}_b)$. The covariances arise because the crossclasses come from the same clusters; they are usually positive and thus they often reduce the variances appreciably. Theory and computation of covariances readily follow those for the variances [Kish 1965, Section 6.5]. However, there may be too many to compute and present: For each crossclass set with k categories there are $k(k - 1)/2$ pairs of differences. A useful rule has been found in numerous computations for large varieties of data:

$$s_c^2/n_c + s_b^2/n_c < \text{var}\,(\bar{y}_c - \bar{y}_b) < \text{var}(\bar{y}_c) + \text{var}(\bar{y}_b). \qquad (2.6.3)$$

That is, the terms 2 $\text{cov}(\bar{y}_c, \bar{y}_b)$ reduce the effects of the pair of crossclass variances for $\text{var}(\bar{y}_c - \bar{y}_b)$, yet these remain greater than the srs variances on the left. Empirical results, though subject to sampling fluctuations, fall generally between the two extremes (each often assumed naively) and more often nearer the srs end. Thus the covariances represent welcome and important reductions of design effects for variances of differences of crossclasses.

Further simplification may be seen by introducing assumptions from (2.6.2) into (2.6.3) and using pooled estimates for s^2 and for roh:

$$s^2\left[\frac{1}{n_c} + \frac{1}{n_b}\right] < \text{var}(\bar{y}_c - \bar{y}_b) < s^2\left[\frac{1}{n_c} + \frac{1}{n_b} + \text{roh}_t\left(\frac{2}{a} - \frac{1}{n_c} - \frac{1}{n_b}\right)\right]. \tag{2.6.4}$$

This simple formula provides narrow limits when the last term is small because both roh_t and the crossclass clusters n_c/a and n_b/a are small. The upper limit on the right is usually a "safe" overestimate because the neglected term $-2\,\text{cov}(\bar{y}_c, \bar{y}_b)$ would probably introduce a greater decrease than the effects of using pooled values of s^2 and roh_t.

Note, however, that perhaps we have been too symmetrical in discussing the variances of $\bar{y}_c$ and $\bar{y}_b$. When n_c (and $\bar{M}_c$) is much greater than n_b (and $\bar{M}_b$), the variance var $(\bar{y}_b)$ will dominate the variance $\text{var}(\bar{y}_c - \bar{y}_b)$. And in that case, $\text{deft}^2(\bar{y}_b)$ may approach 1 and $\text{var}(\bar{y}_c - \bar{y}_b)$ may simply approach s_b^2/n_b.

Some technical, cautionary remarks are in order. To establish simple, useful, "portable" relations based on roh and $\bar{b}$ we had to overlook some complications that may well occur in some designs. The few clarifications that follow may help, but some situations may require more technical treatments [Kish 1965a; Kish and Frankel, 1974; Verma et al. 1980].

1. *Weighting* for unequal selection probabilities can be handled by using the element weights for computing both the actual variances $\text{var}(\bar{y})$ and the s^2, but n in s^2/n remains a simple count of cases. The deft^2 and roh_t will each include the effects both of clustering and of unequal weights. Separating the two effects would introduce more complex computations, as would poststratification or ratio estimation. Note that highly divergent weights may cause great increases in variances and then values of deft^2 will not approach srs variances even for small sample clusters $\bar{b}_c$. *Increases due to inefficient weights tend to be inherited by crossclasses.*
2. *Separating components* of clustering from those of stratification, and each for several stages, would require technical resources that are

seldom used or justified. We depend on the simple and common techniques of "ultimate clusters" or "primary selections."

3. *The cluster sizes* $\bar{b}_c$ were assumed to be fairly equal and not too small. It would be difficult to state useful limits for these vague cautions. In case of doubt about the "portability" of *roh*, it may be checked with more complex computations. See Hansen, Hurwitz, Madow I, Section 6.8, 12D.
4. *Divergent designs* in diverse portions of the sample would interfere with using a single value of $\bar{b}$. For example, large clusters may be used in rural areas, and smaller segments or even single dwellings may be used in cities. Those situations may require separate computations for such divergent design domains.
5. *Overlapping crossclasses* occur when some (or all) of n_c is included in n_b. These situations may need separate treatments that were neglected earlier. However, covariance computations can readily deal with them [Kish 1965a, Section 12.3].

2.7 FOUR OBSTACLES TO REPRESENTATION IN ANALYTIC STUDIES

Many researchers agree to the need for (or, at least, the desirability of) both types of randomizations: randomization of subjects over the population (representation), needed for enumerative (descriptive) studies, and randomization of subjects over treatments, desirable for analytic studies. But many agree to those two needs only separately; then they deny the need for randomized representation in analytical studies.

In my view (1.4) multivariate relations, as well as univariate values, depend on their origins in individual elements, in population bases, and in their environments. The two needs for randomizations over treatments and over populations have similar sources. The two needs are not mutually exclusive, as they are assumed to be in the separate literatures of the designs of experiments and of surveys. The obstacles to randomized treatments, especially in broad populations, are well known (1.3). Here we call attention to four major obstacles to randomized representation for analytic studies. These studies involve generally the search for basic relations between variables; for causal relations from predictors to predictands; for etiological relations of stimulus or treatment to response; for inference and predictions to other populations, to subpopulations, and even to individuals.

First, randomized representations for analytical studies are not common because often they seem too difficult, expensive, impractical, or unfeasible.

These obstacles are many, varied, and only seldom overcome (1.3). Even when feasible, the complex sample designs required often complicate the analysis, as noted under the third obstacle.

Second, they face obstacles and resistance on social, moral, and ethical grounds. The double randomization over populations and over treatments faces questions of "fairness" and "informed consent," especially in experiments with high risks of harm or unequal benefits; and also in "double-blind clinical trials" [Kleinbaum et al. 1982]. It seems much easier to obtain for such experiments and trials the cooperation and understanding of some cohesive group, organization, university, clinic, community, site (3.1A), or several sites for internal replication (3.1B).

Third, mathematical statistical difficulties pose formidable obstacles to handling exactly the double complexities of statistical analyses embedded in complex selections. The covariances of multivariate relations get enmeshed in other covariances (because of complex selection designs) between elements and between other sampling units of multistage sampling. Added to theoretical problems are computational difficulties. Only recently have we seen progress in joint work on theory and on modern computing techniques (7.1). Simple random selection would facilitate the mathematics, but it could greatly increase the difficulties of the first two obstacles.

Fourth, because of those theoretical difficulties, sample survey theory was developed separately from the theory of experimental designs. It was developed to deal with descriptive statistics appropriate for enumerative uses of surveys, mostly for means and totals, and later for differences of means. Today, however, surveys are also being used increasingly for analytical statistics, and theoretical developments are now trying to catch up with practice (7.1).

In addition to those four real obstacles, we also encounter another, which is more artificial, in the denials of the need for representation. Thus we may hear or read that randomized representation, while necessary for descriptive, enumerative surveys, is not needed for analytic studies. Deming has written clearly and eloquently since 1953 on this distinction: "The author distinguishes between enumerative studies and analytic studies. An enumerative study has for its aim an estimate of the number of units of a frame that belongs to a specified class. An analytic study has for its aim a basis for action on the cause-system or the process, in order to improve products in the future" [Deming 1975]. Descriptive and comparative studies are two other names used for this distinction, which reflects closely the divergence between emphasis on representation versus randomization in the literature on surveys versus experiments (1.3). However, I propose that enumerative surveys must also transcend, in aims and in uses, the frame population (2.1). On the other hand, analytical studies must also be concerned with the representativeness of experimental units to the population of inference of the causal system of relations (1.4). Thus I doubt the clear philosophical distinction. But I see practical distinctions. For example, the decrease of the importance of the bias term in Figure 2.4.1 should imply similar decreases for biases of representation.

It is also said that the search for universally valid basic relations should be free of, and not conditional on, the population base of the sample. A similar, if less extreme view, holds that the "internal validity" of the sample is "primary to" (more important than) its "external validity" (3.5). This philosophical basis is similar to the "model-based" approaches of mathematical statisticians and econometricians (1.8), though these are mathematically more sophisticated—and, alas, more difficult to comprehend. Their theoretical basis is stronger mathematically, but not philosophically.

Yet another unwanted obstacle is raised sometimes: What good are statistical validity and sampling errors from probability sampling if we are faced with larger errors and uncertainties due to measurement errors and to nonexperimental designs? This brings us back to our start with the need to compromise among the three criteria of realism, randomization, and representation (1.1). We should not sacrifice representation unless we can gain as much (or more, for the same cost or effort) in measurement or in control of the design. We can view later (7.2F) the compromise broadly as planning the lowest "total error" or "mean square error" for the given situation and cost. Ironically this obstacle is just the reverse of the "attacks on statistics" in 1.8 that call for more observations to reduce sampling errors. However, this obstacle has a weaker base for analytical studies than for descriptive statistics, because the sampling errors are larger and the effects of biases are typically smaller for the former. First, the sampling errors increase with decreased sampling bases for differences, and even more for double differences, etc. (e.g., Was the increase in hypertension greater for blacks than for whites? Was the difference in increases greater in the old than in the young?). Complex analyses quickly get into small cells. Then add the problem of fewer sites (3.1B). Second, the effects of measurement biases are usually less for the more complex comparisons and analyses, as in Figure 2.4.1 (7.2F).

Despite all those practical, mathematical, and philosophical obstacles, we need not become discouraged. On one hand, surveys based on probability sampling are being used for ever deeper and more complex analyses. On the other hand, we must admit happily that even without randomized representation, scientific studies have made great advances, sometimes with treatment randomization in experiments and also without it in observational studies. Medical sciences have made great discoveries with experiments on animals within laboratories and with observations and treatments of humans in clinics. Perhaps in agricultural experiments even more knowledge is needed to apply results from plots and situations in one environment to other populations. In the social sciences most results are bound by culture, society, period, and environment. (Though sociologists are searching and arguing for universal traits.) In all the sciences and in all fields we must struggle to transfer results from one population (or one set of them) to other, by using higher generalizations, scientific theories, and models. The lack of randomized representation, like other defects of induction, noted by Hume's philosophy, will not stop the practical advances of the sciences. Scientists, like philosophers in a quip of Bertrand Russell, often do the right thing without knowing the reason.

It is only rarely that we can achieve randomizations both over a large population and over treatments. We must aim at the feasible and consider strategic compromises and tradeoffs, but keeping both criteria of randomization in mind. And we should neither deny a criterion nor elevate to a virtue those restrictions on it as are forced on us by practical limitations (1.3).

CHAPTER 3

Designs for Comparisons

The Royal Society's motto, Nullins in Verba, *has been translated, "Take nobody's word for it; see for yourself" . . . To be an experiment an experience had to be repeatable. Daniel Boorstin,* The Discoverers.

3.1 SUBSTITUTES FOR PROBABILITY SAMPLING

3.1A Restricted Sites: Community Studies

Confined research sites are often used to restrict samples to convenient and modest-sized populations. They may permit lower costs and easier practicability, especially for longitudinal projects spread over long periods. Good examples of these can be found in the series of health studies based in the cities of Framingham, Massachusetts; Hagerstown, Maryland; and Tecumseh, Michigan. Those projects facilitate studies in depth of the connections and relations between diverse sets of variables, obtained in several studies spread over time. Such rich content also characterized pioneering social research sites, such as the Middletown of the Lynds in Muncie, Indiana; the early fertility studies of Indianapolis; and the long series of Detroit Area Studies. Much knowledge and insight have been gathered from those community studies, which were later applied on broader, national scales. Furthermore, a "site" in this discussion may also refer to a defined social organization—university, business firm, social club, labor union, etc.—as well as a community or local area.

Attempts are commonly made to choose sites that are not only convenient and economical, but also somehow "typical," sites that incorporate in "microcosm" a larger, perhaps national population. Such hopes would be more justified if that microcosm actually consisted of the equivalent of a representative (or random?) selection of elements from the larger, target population. But the bases for such hopes are not firm or broad, and the gaps of inference to the target population (2.1) can become much greater than those for widespread or national samples. Nevertheless restricted research sites and studies of single communities can be justified on grounds of costs, economy, and feasibility, especially in evaluation studies (3.7).

Those practical grounds for severe restrictions should not, however, be confused with attempts to make virtue out of that necessity, that is, with deliberate and drastic restrictions of studies to "homogeneous populations" or to "pure types" of elements (3.1C5). Such narrow restrictions—seeking to exclude perhaps some disturbing variables from the larger population—can seldom be justified on statistical grounds (see quote from Fisher in 1.3). The best statistical defenses of such restricted "monograph" studies date from around 1900, before the arrival of modern theories of randomization in surveys and experiments [O'Muircheartaigh and Wong 1981].

We may admit that restricted sites may be clearly defined and often frames constructed for their entire populations readily obtained. However, inferences from them to larger, more meaningful national populations are much more difficult and tenuous (2.1). The dilemma can be highlighted by a complete coverage (census) from a restricted site: Few researchers, if any, would depend on complete coverage in order to forgo sampling errors, confidence intervals, and significance tests. Yet those probability statements acquire meaning only in the framework of some population or universe from which the sample elements can be supposed to have been selected. However, inferences to broader populations require models (1.8), and for bridging the gap from models to desired populations the restrictions of sites pose several handicaps. Local sites sacrifice broad (national) representation in space for the gains of depth in time, in richer variables, and in relations that can be discovered more fully and economically. Sometimes this trade-off is acknowledged by counterposing "longitudinal" to "cross-sectional" studies. However, we must guard against the confusion between the two dimensions implicit in that contrast: Longitudinal studies sacrifice representation over space (the national target population) for better treatment over time, whereas single, cross-sectional, national studies sacrifice depth in time. Longitudinal studies of national cross sections can represent both the spatial and time dimensions; and neither dimension is covered in a one-time study on a restricted site (6.1–6.2).

Furthermore, restricted sites also suffer from changing boundaries, and from "internal migration" of movers over local boundaries. Hence a major purpose of longitudinal studies, research on individual (micro) changes, tends to be frustrated. To follow movers beyond local boundaries becomes increasingly expensive over time, wasting some of the savings that motivate the use of restricted sites. On the other hand, if the study is confined to nonmovers, it will become increasingly biased. Hence longitudinal studies in restricted sites generally tend to be concerned only with mean (i.e., net, macro-) changes. Furthermore, mean changes must also depend on assumptions that the balance of immigration and outmigration reflects and is somehow representative of changes in the target (national) population, and that may be only wishful thinking.

Single organizations as research sites present similar problems and concerns. A seemingly stable, fixed social unit suffers in- and outmigration of its people; this is true of universities, business firms, households, etc. The identity of organizations and communities over time is discussed in 6.3.

Probability selection within sites does not fill the gap of inference from the sites to the target populations or to the populations of inference, but it does fill another need for objectivity. Therefore probability selections within sites should be pursued, and they can often be achieved more readily and economically than for selections on a national scale. This advice goes for designs confined to small numbers of sites in 3.1.B.

3.1B Internal Replications: Several Sites

Early in the planning of many research designs the limitations of a single site must be balanced against the costs of a national sample. The need for many sites (cities, counties, etc.) for national samples raises obstacles as frequently as the need for many individuals. Important results from single sites in medical research are quickly tested with replications on other sites because there are so many medical researchers. Such *external replications* are useful both when they tend to confirm valid new results and when they contradict them, raising cautions about early and dubious new results. For other examples, see the contradictions between election polls (7.6). But in social research external replications are usually more difficult, because variable conditions, situations, etc., make it more difficult either to confirm or to contradict "similar" results and because, alas, replications are not fashionable. Instead we should look to *internal replications within single research designs*.

Internal replications offer useful compromises between a single site and a national sample that could require, say, 100 sites, and they can serve diverse purposes (3.1C). A compromise solution may have 4 or 8 or even 16 sites, for example. The design may also call for pairing "treatments" with "controls" either within each site or with matched sites. For k replications, matching within sites would need k sets of sites, each to contain both treatments and "controls." Sometimes this design is not feasible because the treatments must involve entire sites; then k replications of treatment plus control require $2k$ sites. Furthermore, for unbiased error computations based on random replicates (to remove the between-sites components from the comparisons), we would need $4k$ sites for two random duplicates and $2mk$ sites for m random replicates.

With few sites it would be vain to attempt representation with probability sampling of a population (1.4). Nevertheless we should try for wide spreads across important disturbing variables to obtain wider confirmation. Here we may apply a modern philosophical view: The choice of the sites should strain to increase the possibilities for "falsification" (7.6) [Popper 1959, Salmon 1967, Magee 1973]. Thus four sites allow for representing highs and lows or treatments and controls for two variables; thus with A/a and B/b representing two treatments for two variables, the four sites will have *AB–aB–Ab–ab*. With "confounding" in "Latin squares" four sites can represent three variables, each balanced against the other two variables: *ABC–aBc–Abc–abC*. With nine sites three treatments, for each of four variables, can be accommodated by "Graeco–Latin squares." We need not and cannot repeat here the advantages (and problems) of multifactor methods that have been richly developed for experimental designs [Cochran and Cox 1957; Cox 1958; Snedecor 1967; Anderson and Bancroft 1952].

Similar and consistent results from the replications yield stronger confirmation than a single site would. But if the results are discordant, the replications are too few to yield dependable inference; then further research is needed. Nevertheless, discordant results still yield a healthy skepticism that naive "success" from a single site would have obscured.

3.1C Designs for Internal Replications

We need to face questions of design concerned with the field of representation intended for a few sites. Should the sites represent strata of the entire population, and should they be randomly selected? Or should the sites represent extremes of the variables, purposely chosen?

1. ***Representative Domains.*** Probability sampling requires large numbers of sampling units, of sites as well as of elements. But when we are confined to a few sites we must depend to a considerably greater extent on model-based assumptions. The closest design to a probability sample would be its imitation: *dividing the entire population into a few different domains* and then choosing within each domain a site, or a pair, or a set of sites. Thus the samples and the comparisons would be designed over diverse portions of the entire population, and the internal replications would be conducted over the full range of the population distribution. Three decisions seem most important here.

First, how many sites can we afford altogether? Within that total, the

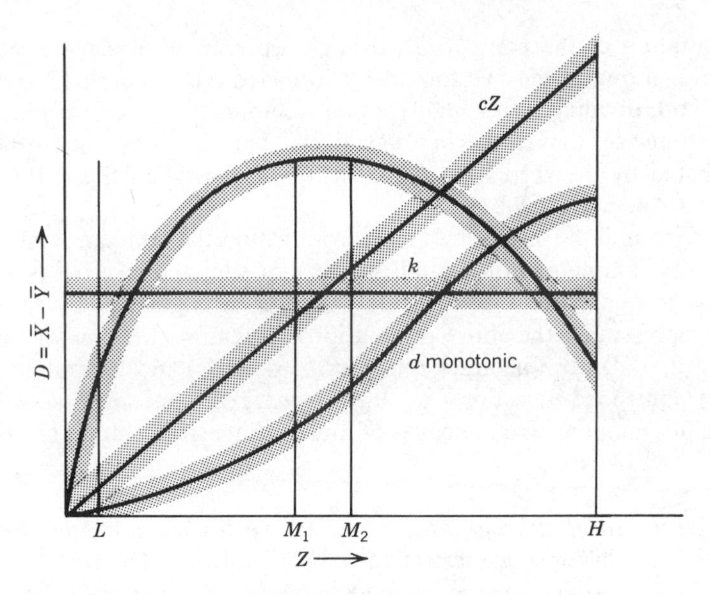

Figure 3.1.1. Selecting a single or a few sites to measure $D = (\bar{X} - \bar{Y})$.

When the effect $D = \bar{X} - \bar{Y}$ is constant (k) over the control variable Z, any value (L, M, or H) of Z will tend to yield similar values (k) of D. Yet having two (or more) "modal" or "typical" sites (M_1 and M_2) would tend to reveal the "random" variation in d between sites, as symbolized by the bands around the expected values (k). The statistics d'measured at each site will also be subject to sampling variation within sites in measuring d. The constant expected value (k) corresponds to a simple "additive random" model in experimental design.

To discover a linear or monotonic relation of D to Z, at least two "extreme" sites, low and high (L and H), are needed; and at last two sites at each extreme to separate the variation in site values d from the functional relation of D. With a strong linear model, the slope of the regression line of D on Z may also be estimated. The mean of sites values may also be a "fair" estimate of the average effect $\bar{D}$, which may be a principal aim of the study. This may also be strengthened with three (L, M, H) sites or more, selected so as to represent different values of the disturbing variable(s) Z, and such designs may also discover relations that are not linear, or not even monotonic.

Another extension should consider variations in D with more than one variable Z, but this would be more difficult to represent on this flat page. Such considerations lead to L/H designs in 2, 3, 4, or more dimensions of $2 \times 2 \times \ldots$ designs.

number of sites per domain, hence the number of domains, must be decided (3.1D).

Second, how should the population be divided into separate domains? This may pose dilemmas because the distributions of relevant stratifying variables are usually not even and smooth: Their meaningful cutting points may be located near one or both ends of a continuum, or even at several places along a line. Dividing near the extremes can result in loss of representation of the population; though representation can be recovered with reweighting, it may be too costly in increased variances (7.4). On the other hand, dividing into roughly equal domains may sacrifice meaningful distributions of the stratifying variables. This dilemma often becomes exacerbated by the need for several stratifying variables [Kish 1965a, 3.6; Kish and Anderson 1978].

Third, should the selection of sites from within the domains be done with probability sampling? For a small number of sites the case for probability selection is weak. But that case may become strengthened by the desires (a) for representing the entire population, as against the aims of the other designs (2, 3, 4); (b) for larger numbers of sites; (c) and for public presentation. Controlled selections or Latin-square designs may satisfy both probability sampling and the needs of multivariate stratification (7.2F) [Kish 1965a, 12.8].

2. ***Typical, Modal, Average Sites.*** In 3.1A we discussed the frequent use of and justifications for single restricted sites for research. The reader probably can find other good examples from experience and reading; hence we need not elaborate on them. The choice, deliberate and judgmental, of single sites becomes in most situations aimed at a "typical," hence modal or average, unit (city, community, school, etc.). It comes as a disappointment sometimes that statistics and sampling can contribute so little technically to the choice of single sites; yet an experienced statistician and sampler can help with a creative, philosophical dialogue. The difficulties may be illustrated by trying to agree on an "average city" in the United States—clearly not New York, Boston, Philadelphia, Washington, Miami, Los Angeles, San Francisco, etc. Is Muncie, Indiana, "modal," even though more people live in "places like New York"?

An average or modal site is chosen to serve as "representative" of conditions related to the study, although that aim may be veiled modestly with the double negative of "not unrepresentative" or even with a false disclaimer like, "The single site cannot be representative, yet. . . ." One should accept the site as the target population, then face the chasm to the desired inferential populations (2.1) and describe a model for the meaning of "representative" (1.7, 1.8).

Several "average" sites may be chosen for internal replications and a pair or several pairs may be chosen for treatment/control comparisons, or k-tuples for k treatments. The choice of average sites usually involves "balancing" along several control variables.

3. ***Contrasting, Extreme Replications.*** Instead of one or more average sites, the aims of "falsifiability" seem better served with tests obtained in contrasting conditions (3.1B, Fig. 3.1.1, 7.6). When we can assume a possibly linear, or merely monotonic, relation from the stratifier (control) variables to the research (explanatory) variables, we can gain a more robust test of falsifiability and a stronger confirmation from extreme units than from intermediate or "representative units." For models of linear regressions the image is clear and well known: The end points of the regression line provide optimal allocations for sample points to measure with highest reliability the slope of the linear regression. High versus low extremes yield better tests than would "representative" selections obtained within a complete division of the population into high and low halves, as in 1 above. The extremes are more informative than either random or modal or centralized selections from the halves would be. This is also true for monotonic relations generally, not only for linear regressions. The extremes also provide more informative tests than would one or two average or modal selections, as in 2 above.

If both high and low units (sites) yield similar results for treatment/control comparisons, we gain indications of no effects from the stratifier: a flat slope for the regression line. On the other hand, differences between the effects indicate a slope for a linear or monotonic relationship.

For less reliance on linear (or monotonic) relations we may want to sample more than the two extreme points (units, sites), and a third unit (site) near the center may well appear useful. Effects near the center that approximate an average of the two extremes would tend to indicate linearity (or at least monotonicity). On the other hand, divergent and contradictory results may indicate a curvilinear or complex relation, the need for more sites to identify and quantify the relationship, for better and more control of disturbing variables, and perhaps for probability sampling of the entire population. Probability sampling on many sites should be designed for the estimation of average effects over the entire populations of inference.

4. *Distinct, Divergent Replications.* Distinct, divergent replication may be designed specifically to submit theories (hypotheses, conjectures, relations, treatments, comparisons) to the most difficult tests of falsification—but perhaps without any specified model of linear or monotonic relationship. This merely restates the aims of internal replications (3.1B): Using only one site or a few sites of one specific type would lead us to the meager results of 3.1A, and good statistical design calls for diversification over disturbing variables. A large number of sites would permit a complex multifactor (factorial) experimental design or a probability sample over the population and its major domains.

5. *Experimental Designs.* Dozens of books on the subject of experimental design are available at different levels of depth and difficulty, applied to most fields of research. Most books are longer than mine, and I cannot find a sensible way to broach the subject briefly. The designs of experiments have also been applied to observational studies. Representation of the population is difficult in experiments, as we noted. But the multiple controls on few sites that "Latin-squares" permit have their analogy in survey sampling in "multiple stratification" (controlled selection, deep stratification) (7.1.D) [Kish 1965, 12.8; Cochran 1977, 5A.5].

3.1D Number of Sites

We assume here that the total number s of sites (units) must be narrowly limited because of high costs per site. Thus we need to discuss briefly the relations between the numbers of treatments t, the numbers of domains d, and the numbers of replicates r for each treatment within domains. With the total number number $s = t \times d \times r$ of sites severely limited, the more we take of one component, the less we get of others. Instead of symmetrical designs we may have different numbers of replicates, domains, and treatments, represented by the product $\Pi t_i d_h r_j = s$. The number of treatments can be as low as 1, or 2 for treatment plus control. The number of domains was discussed briefly in 3.1C; we focus here on the numbers of replicates for treatments r_t and for controls r_c. The number of observations (individuals, elements) at each site is also important, but that subject can be treated better separately in connection with the kind of analyses, both substantive and statistical, planned for the data. Some alternative designs are sketched below; based on technical considerations, their application should depend on resources, situations, and models that can differ greatly.

Match d treatments with more controls ($d \times r_c$) when controls are less expensive. This needs $d(r_c + 1)$ sites for d domains. Rigidly symmetrical assumptions have restricted many studies to d pairs of treatment/control comparisons, even when much lower costs would allow for many more controls. Several (r_c) controls in each domain reduce the errors of the controls, and they also allow the computation of between-site errors with $d(r_c - 1)$ degrees of freedom. Here we may also need to overcome a common inertia for choosing as controls those sites that seem "closest" (most similar) to the treatment sites; however, in most situations several (or many) sites may be almost as close. Perhaps construct domains of ($r_c + 1$) "similar" sites; choose (at random) one site for treatment and use the other r_c sites for controls. For example, if a population of 420 sites is sorted into $d = 4$ domains of 105 sites each, each treatment from the four domains may be matched with $r = 10$ controls in a total of $4(10 + 1) = 44$ sites.

Consider r_t treatment sites and r_c control sites in each domain. This would permit reduction of errors for both treatments and controls; also the measurement of between-sites errors for both, with $(r_t - 1)$ and $(r_c - 1)$ degrees of freedom. This must be balanced against needs for more domains, and for more than two treatments, as well as against limits on total number of sites. Consider $r_t = 2$ for a design with a total of $(2 + r_c)d$ sites. In the preceding example we could have $r_t = 2$ treatment sites and $r_c = 10$ control sites for a total of eight treatment and 40 control sites in the four domains.

Consider treatments and controls from the same sites, if the applications of treatments and of measurement permit. This should yield good gains, both by allowing for more domains for both (or all) treatments and by eliminating or reducing the effects of between-site variation from the comparisons. Within the sites, large numbers of sampling units often may be used efficiently. On the contrary, when separate sites are needed for each treatment and control, the number of sites becomes too large for several variables d. Thus for treatment/control pairs we need $2^2 = 4$ sites for high/low domain for one variable; and $2^d \cdot 2 = 2^{d+1}$ sites for high/low and treatment/control pairs for d domain variables. For t treatments and r replicates we would need $tr2^d$ for high/low contrasts for d variables. However, these numbers can be reduced with Latin-square and "fractional factorial" experimental designs.

3.2 BASIC MODULES FOR COMPARISONS

3.2A Principal Features of the Modules

Controlled observations and quasi-experimental designs have a wide but diffuse literature in the social and medical sciences. That diffuseness contrasts with the two distinct and compact literatures for experimental designs and for sample surveys. Each of these can present basic tools in a compact introductory treatment, but only by narrowly limiting its objectives and by using strong assumptions. Each field made fast progress in 50 years by traveling on a well-paved narrow avenue. However, the literature of controlled observations is far more diffuse, because it reflects the varied and amorphous nature of the area it covers.

I believe this subject also needs an introduction that is both general and compact, so that it can be taught and learned. With the following design modules I aim to encompass this diffuse, general area in a simple, compact presentation and with a minimal set of clear symbols. These modules will let us build (in 3.4) the more complex research designs that are commonly used. Furthermore, this method of modular design facilitates the evaluation of those designs. Our treatment should be comprehensive and include all basic aspects of good design in order to be useful in practice. Such a comprehensive treatment of all aspects, even if superficial, is more useful for an introduction than the typical thorough treatment in depth of just a single aspect. Essentially the simple modules combine the variance components of experimental designs with cost components and then join them with the bias components associated with observational studies; then the modules become building blocks for study designs. The following basic features of these four modules use drastic simplifications that are necessary to encompass *simply* the diverse aspects of designs and to permit comparisons.

1. *Four modules are defined in terms of their mean square errors*: MSE = variance + bias^2, for assessing the *relative accuracies* of the modules. With each module we associate specified factors of unit variances and unit costs, as well as specified major sources of bias. This joint comprehension of variances, costs, and biases makes these modules useful as building units for comparative analyses of complex designs.
2. *Research designs are built from the four modules.* The building process is exemplified by the five most common basic research designs in Section 3.4. Other and more complex designs can also be built from more combinations, and from modifications of these modules. The relative problems and the advantages of the complex research designs can be readily assessed in terms of their simple component modules.
3. *Bias is specified in four major types.* This intermediate step facilitates the introduction of the 22 classes of actually recognizable forms of bias later (3.5). Those types and these classes serve to bridge the vast gap between a single, general, unspecified bias term and the myriad specific actual biases one can conceive and sometimes even measure. Occurrences of the major types differ between modules, therefore also between designs; thus those differences facilitate comparisons of bias types between designs.
4. *Research designs are distinguished by clear patterns of variances, costs, and bias types.* These follow from the modules used for constructing the research designs and from the clear association of variance, cost, and bias terms with the modules. Most bias terms concern only controlled observations and vanish from the corresponding designs of randomized experiments.
5. *The modules, symbols, and terms are related to and clarify the literature on controlled observations.* These follow the concepts for designs and for sources of bias used by Campbell [1957, 1963] and by others before and after him. However, that literature concentrates on biases only, whereas I introduce variances and costs also. They can also be related to the concepts of experimental designs; but these typically concentrate on variances and neglect the presence of biases of nonrandomization. Links to the concepts of survey sampling are also noted below (3.2B).

6. *Variances and costs are presented in simple relative terms for comparing research designs.* Complicated nuisance parameters are isolated and postponed so that the modules and the designs can be compared with only two parameters: V_i for variances and k_i for costs as factors specific to the modules *i*. Variance components are $V_i S^2/n$ and cost components are cnk_i; the $V_i k_i$ (and four bias symbols) are used to compare modules and designs, because the n, S^2, and c cancel in relative comparisons. This economy of symbols permits wide generality yet simplicity at the same time. The postponed complications of practical situations can be reintroduced later in the 22 classes of biases (3.5) and in enlargements on the limitations (7.2).

3.2B Limitations of the Modules and Symbols

For simplicity in the presentation and comparisons of modules and designs some limitations had to be introduced. These limitations occur in the literature either implicitly or hidden by unreal assumptions, but I prefer to expose them openly, even if briefly here. We shall later (7.2) consider methods for relaxing all these limitations in order to serve with greater generality more realistic and complex designs.

1. *Differences of pairs of means* $(\bar{x} - \bar{y})$ *are used for the comparisons.* These are the simplest and the most common forms used for comparing treatments in controlled observations, in experimental designs, and in sample surveys; for example, for comparisons of A–B treatment means, with–without means, subclass means, before–after means, etc. Furthermore, the results for $(\bar{x} - \bar{y})$ we develop here to compare research designs can also be readily applied, with only simple modifications, to other comparisons: to ratios of means; to ratios of ratios; to differences among several means; etc.
2. *Equal sample sizes n are used for the two means.* This kind of symmetrical design seems simple and common; perhaps they are too common, because departure from equality and symmetry may be desirable (3.1D). Our modules can be readily modified to fit the different sizes.
3. *Simple random selection is implicit in the variances* S^2/n *for the means.* This assumption underlies accepting S^2 as the variance of elements in the population. But that limitation can be overcome by introducing design effects D_i^2 to modify the variances, and then we can have $S^2 = D_i^2 S_i^2$. Thus the variations in element variances S^2 and design effects D_i^2 can be considered separately. This scheme can also facilitate the treatment of more complex and different cost factors c_i.

4. *Representation of the population(s) with probability sampling is treated separately* (1.7, 1.8). It would detract from the simplicity of the symbols for modules to burden them with this task also; a separate treatment better serves our aims. But four sources of bias for representing populations are noted among the 22 sources later (3.5).
5. *The five basic research designs considered here (3.4) are the simplest, most used, and most practical.* But the modules and the methods can also be expanded to more complex research designs.
6. *Mean squared errors are used for a combined view of biases with variance and cost factors.* They perform rather well when biases can be assessed approximately in relation to the sampling variance and when the distribution of the means is approximately normal. This is the most common and useful approach to combining these two distinct types of errors, but other approaches are theoretically conceivable.

3.2C An Example of Relative Cost Factors

This example precedes the formal development in 3.3 of factors and modules for those of us who can more readily perceive a specific, numerical example and then follow better the abstract presentation. Others may postpone this to the end of 3.3.

Cost factors are symbolically combined with variances, so that these may be used to compare diverse designs for *any fixed aggregate cost.* The factors are fixed symbolically only so that the comparisons are valid for varying levels of actual expenditures. The basic cost factor, fixed at $k_1 = 1$ for convenient comparisons, is for the indispensable core module, which is common to all the designs: It contains n experimental elements, and its cost is cn, where $c = ck_1$ is the element cost of the "experimental" (treatment) observations. The other factors k_i are made relative to the basic, common factor $k_1 = 1$.

Next, for example, consider a design based on n treatment plus n control elements for an aggregate cost of $cn + c_3n = cn[1 + k_3]$, where $c_3 = ck_3$ is the element cost for the "controls." The cost factor for introducing the controls is $[1 + k_3]$, where $c_3 = ck_3$. Thus in comparison with the variance S^2/n for the basic design, the variance of a "control group comparison" is seen as Var (design 3) $= 2(S^2/n)[1 + k_3]$; to keep the total cost constant in the comparison, the sample size is decreased, hence the variance increased, by the factor $[1 + k_3]$. When the controls cost as much per element as the treatments $k_3 = 1$, and the introduction of controls is seen to increase the variance, for fixed total cost, by $2[1 + k_3] = 2 \times 2 = 4$. When controls are free, $k_3 = 0$, and controls increase the variance by a factor of 2 only.

Generally the element cost for controls is not greater than for treatments, but they cost something, thus $0 \leq k_3 \leq 1$.

For example, suppose that $c = ck_1 = \$100$ for experimental treatment plus testing the aftereffects per subject included (and persuaded to participate) in the experiment. Suppose also that the control cases cost somewhat less: $ck_3 = \$60$ per subject in the control, because \$40 represents the extra cost of the experimental treatment. Then $k_3 = 60/100 = 0.6$ and $[1 + k_3] = 1.6$. Thus the variance of design 3 is increased by $2[1 + k_3] = 3.2$ over the basic design 1. Hence for a *fixed total cost of \$32,000* we can buy (200 treatments × \$100) plus (200 controls × \$60) = \$32,000, and the variance of 200 pairs is $2S^2/200 = S^2/100$. However, with design 1 we could afford 320 subjects, since 320 × \$100 = \$32,000; and its variance would be $S^2/320$. Thus the variance of design 3 is 320/100 = 3.2 greater for the same fixed cost. Alternatively (but less realistically), one may insist on a desired *fixed variance*; for example, on a desired $S^2/100$ for the difference of the two means. Design 1 would obtain this for 100 × \$100 = \$10,000, whereas design 3 would need the \$32,000, as earlier and the contrast of the two costs for a fixed variance is again in the ratio of 3.2.

Thus either for fixed costs or for fixed variances this modular presentation, in terms of unit cost × unit variance, can readily contrast the *relative* efficiencies of research designs. In the presence of *estimated biases* we may use the mean squared errors to also include estimates for biases squared along with the variances.

3.3 FOUR MODULES: COSTS, VARIANCES, BIAS SOURCES

3.3A Costs and Variances

These four modules serve as building blocks for research designs; first and foremost for the five most common basic designs in 3.4, but potentially for other and more complex designs as well. We need symbols for these modules that will be (1) simple, (2) descriptive of their functions, and (3) in tune with the literature on this subject. I propose the following symbols for the modules.

I. $[.Ex]$ = Experimental treatments (E) followed by observations (x).

II. $[XE\tilde{x}]$ = Experimental treatments (E) with pretest (X) and posttests ($\tilde{x}$).

III. $[.Ob]$ = Controls (O) followed by observations (b).

IV. $[BO\tilde{b}]$ = Controls (O) with pretests (B) and posttests ($\bar{b}$).
IIIR. $[.\text{Ø}c]$ = Randomized controls (Ø) followed by observations (c).
IVR. $[C\text{Ø}\tilde{c}]$ = Randomized controls (Ø) with pretests (C) and posttests ($\tilde{c}$).

E denotes the experimental treatment and O the control, which may be none, or standard, or placebo. But we distinguish with Ø research designs when true randomization is used to separate the experimental subjects from the control subjects. In the experimental group, X denotes observations before treatment and (.) its absence; and $\tilde{x}$ or x distinguish observations after treatment in the two cases. The corresponding symbols in the control group are B or (.) for with/without observations before control treatments, and $\tilde{b}$ or b distinguish observations after control treatment in the two cases. These are C and $\tilde{c}$ for randomized controls (Table 3.3.1).

Pretest and posttest are terms used in the literature to denote pretreatment (or precontrol) observations and posttreatment observations, respectively. Thus I shall also use those brief and common terms, despite their inaccuracy: The pre- and the post-actually denote the timing of the observations (tests) before and after the administration of treatments, not of tests. Furthermore, *observations* is a more adequate and general term than *tests*, which comes from educational research. The word *pretest* is also confusing because in survey literature it refers to small samples for trying out methods, especially to field trials of questionnaires, sometimes also called "pilot studies."

The pretest and the posttest may be similar or they may differ. The pretest may be simple and spare, or it may consist of elaborate multivariate measurements. Posttests may vary even more, and they may denote a whole series of observations designed to search for longer-range effects of the treatments (3.6). Treatments also may be complex, drawn out, and repeated.

The sequence Ex denotes experimental treatment followed by observation. This is the fundamental, basic unit present in *all* research designs. Because of that omnipresence we use its cost as the *basic unit cost* ck_1, so that all other unit costs can be treated as *relative* to this basic cost factor with $k_1 = 1$. I created this convenient symbol to serve for comparing the *relative* unit costs of other modules, and later of research designs. The basic element cost $c = ck_1$ denotes the unit cost for each of the n elements, and for n elements the costs come to cn. This is also the cost of the basic module I, symbolized by $[.Ex]$. The actual composition of ck_1 is complex: cnk_1 must include the selection and recruitment of the n subjects plus the experimental treatments plus the posttreatment observations. I believe this complexity is worthwhile because it helps to simplify the comparisons with the other modules, and later the comparisons among research designs. Of course it can be decomposed into its components when the occasion warrants the effort. Its actual value may have to be thus composed from such components.

TABLE 3.3.1. Components of Biases, Variances, and Costs for Four Modules

Module Symbol	I [.*Ex*]	II [*XEx̃*]	III [.*Ob*]	IV [*BOb̃*]
Expectation of Bias	$M + E_t + T$	$E_t + T + P_t + PE_t$	$M + T$	$T + P_t$
Unit Variance	S^2	$2S^2(1 - R)$	S^2	$2S^2(1 - R)$
Unit Cost	$ck_1 = c$	$c(1 + k_2)$	$c(k_3)$	$c(k_3 + k_4)$
Relative Cost	$k_1 = 1$	$0 \leftarrow k_2 < 1$	$0 < k_3 < 1$	$k_4 = k_2$
Cost × Variance	cS^2	$2cS^2(1 - R)(1 + k_2)$	cS^2k_3	$2cS^2(1 - R)(k_3 + k_4)$
Present in Designs	1, 3, 5	2, 4, 5	3, 5	4, 5

The cost of module II, $[XE\bar{x}]$, includes the cost ck_2 of pretests in addition to the basic unit cost of ck_1. The cost for n elements for module II is therefore $cn[k_2 + 1]$. The unit cost ck_2 of pretests should be less than the basic cost ck_1, which includes, in addition to the postobservation, the cost of selection and of experimental treatment. Hence $0 < k_2 < 1$ expresses wide boundaries within which the researcher should try for a useful estimate. If the pretests are very cheap, as a ratio of k_1 with its expensive selection and treatment plus posttests, k_2 may be near zero. Values of k_2 near 1 would arise only when the observations X and $\tilde{x}$ are equally expensive, whereas selection and treatment are relatively very cheap.

Module III, $[.Ob]$, involves selecting the n elements and obtaining cooperation from and observations on them, with unit cost ck_3, compared with ck_1 for Module I. The control treatment (O) may be less costly than the experimental treatment (E). We should search for k_3 within the likely boundaries of $0 < k_3 < 1$. Values near 1 denote equal unit costs for control and experimental treatments, at least compared to costs of selection and observation. Values of $k_3 > 1$ would mean that the control is more expensive than the treatment in k_1, and this is unlikely but possible. Values near 0 would denote the predominance of the cost of experimental treatments over all other costs.

Module IV, $[BO\tilde{b}]$, adds the cost ck_4 of pretests to the cost ck_3 of Module III, and for n observations we have $cn[k_4 + k_3]$. This factor $[k_4 + k_3]$ resembles $[k_2 + 1]$ for module II, and it is likely that $ck_4 = ck_2$ and $k_4 = k_2$, the additional cost of a pretest in each case. Putting it all together for completeness, we suggest that commonly $0 < k_2 = k_4 < k_3 < 1$, and that often we may be able to obtain or guess numerical values for the inequalities, which will be useful though inaccurate.

It seems difficult to estimate the *additional cost of randomized treatments* in modules IIIR and IVR. Were we to force the cost of randomization into the four k_i cost factors, we should include it in k_3 because the cost of randomization arises only with the introduction of control treatments. In that case, instances of $k_3 > 1$ could easily occur. But I think that treatment randomization does not fit k_3 because it is not a unit cost proportional to the number of cases. It seems closer to a fixed cost, independent of n, which can be added as (RdT) to the research design.

It would be even less feasible to include *randomized selection over the population* in the k_2 and I propose a separate constant (RdS) for that. Furthermore, it seems likely that often the two constants will not be independent, that joint randomization both over treatments and over populations would be more costly than just the sum of the two costs. That is why researchers choose either experiments or survey sampling, but seldom both (1.1).

Variance factors are best viewed together with cost factors. Modules I and III each concern the simple element variance S^2. This is the basic unit variance for the modules and the variance of each mean is S^2/n. In a "control group" design (3) we take the difference $[.Ex] - [.Ob]$ of the two means; hence the variance of that difference is the sum $2S^2/n$ for the variances of two independent means. The cost for the design comes to $cn + cnk_3 = cn[1 + k_3]$. For example, suppose the basic unit cost of $[.Ex]$ is $ck_1 = c = \$100$, and the unit cost of $[.Ob]$ is $ck_3 = \$60$; then $k_3 = 0.60$ and $[1 + k_3] = 1.60$. However, in contrast with the "one-shot" design $[.Ex]$, we must note that the "control group" design $[.Ex] - [.Ob]$ also increased the variance by a factor of 2. The joint increase of the variance and cost factors has the ratio of $2S^2[1 + k_3]/S^2[1] = 2[1 + k_3] = 2[1.6] = 3.2$. This represents, regardless of sample size n, the statistical efficiency: either relative variances for fixed cost or relative costs for fixed variance (as in 3.2C). This explains our use of the products of unit variance with unit cost factors for assessing relative statistical efficiency of modules and research designs.

Modules II and IV involve the difference between the pretest and posttest observations. Therefore the basic unit variance for these modules becomes $[S^2 + S^2 - 2RS^2] = 2(1 - R)S^2$, where R is the correlation coefficient between pretest and posttest observation on the same elements. Compared with the basic unit variance of S^2 for modules I and III we note here a factor of $2(1 - R)$. This denotes a decrease in the basic unit variance when the correlation $R > 0.5$. Since correlations on the same elements are often high, taking pretests can often decrease unit variances. However, we must also consider the added unit costs of the pretests, which are in the ratio of $(1 + k_2)/(1)$ for module II and $(k_3 + k_4)/(k_3)$ for module IV. Taking the products of the basic unit variance and cost factors, we note the ratio of $2(1 - R)(1 + k_2)$ for the pretested experimental treatments with module II compared with module I; and $2(1 - R)(k_3 + k_4)/(k_3)$ for control treatments with module IV over module III. Again these contrasts of statistical efficiency hold (regardless of the sample sizes n) both for variances with fixed costs and for costs with fixed variances. *Statistical efficiency* is the term I use to denote jointly the effects of unit costs and variances, but omitting biases. Thus we postponed considerations of biases in the mean square errors.

3.3B Biases: Four Major Types

Now we come to our major and most difficult problem: a symbolic representation for the major types of biases, so that we may consider them jointly with variances in mean square errors = bias^2 + variance. To write merely bias^2 is too distant from the specific sources of the biases in specific studies. We must choose a convenient level on the abstract/specific continuum to retain simplicity yet to seek usefulness too. Five levels of specification descending from the most abstract to the most specific can be designated.

I. Bias in general
II. Four major types of bias: E, M, T, P and their interactions
III. 22 classified sources of bias in all fields
IV. Types of bias specific within fields of study
V. Specific biases for specific studies

For present purposes level II seems the most useful; it creates distinct patterns for the four modules (Table 3.3.1) and for research designs. Thus that level is effective yet simple, since dealing with four major types plus their interactions is not too complicated, and it also resembles some other treatments (Ross and Smith 1968; Namboodiri 1970]. Later (3.5) we can trace these four major types plus two others to the 22 classified sources of bias at level III. These in turn are still general enough to serve any field of research, yet specific enough to direct the researcher's attention to specific problems. These 22 sources on level III follow, in a modified form, the "twelve threats to validity" in the established literature [Campbell 1957; Campbell and Stanley 1963; Cook and Campbell 1979].

Level IV could be even more useful for any specific field of study and for specific techniques of research. For that level of specialization one can describe specific types and values of biases based on actual experiences and discuss reasonable expectations. Further down at level V one would discuss biases based on the actual working grounds of specific studies. However, to cover a wider area we must operate on the more general yet descriptive level III, but begin at level II with the four major types. We must understand that these bias types, E, M, T, and P, represent "expected" values, i.e., averages in the population, and that values for individual subjects are subject to variation. Even their values for sample means can be assessed only with sampling errors, if at all.

E denotes the experimental treatment as a net effect in contrast to *0* effect for the "control." It is not properly a source of "bias" in the usual sense, but it must compete for measurement with the biases, the disturbing factors (1.2). Hence, for dealing with our symbols, E must be listed among the biasing effects. This brief symbol E may oversimplify what actually occurs in many situations, and we may consider $E = E_i - E_j$ as the difference of two distinct treatments. In any case, E and $E_t = E + ET$ symbolize the separation of the predictand variable from the disturbing variables, which are denoted by the bias sources *M*, *T*, and *P*.

M refers to the group mean *before* the treatments *E* or *0* at the time of the pretest observation, which we denote with *X* or *B*, or with (.) if without pretest. This is a source of possible bias $(M_x - M_b)$ for nonrandomized designs, but for randomized experiments $(M_x - M_b)$ should average to zero in expectation. Even when omitted for brevity, the subscripts are implicitly present for the designs. We assume the same values of M_x and M_b for posttests as for pretests, because T_x and T_b are used to carry the changes over time between them.

T denotes effects of disturbing variables (history, maturation, mortality, instrumentation in 3.5) that occur in, and depend on, the time elapsed between pretest and posttest observations. We may also consider $(T_x - T_b)$ for denoting the differential effects of elapsed time for nonrandomized comparisons. This explicit introduction of time effects is an important contribution of our model to elementary statistical presentations, because it may often be of practical importance in research.

P denotes the effects of pretests on subjects. In educational psychology we read of traditional concern with these effects, either as direct learning experience or as a factor that motivates learning behavior. I include it to be consistent with that literature and for caution, but I believe it usually to be negligible in comparison with the effects of *E*, *M*, and *T*.

Each module results in a distinct set of bias terms; then, by building the research designs from these modules, we can associate these designs with those distinct bias terms (3.4). Those bias terms lead us in turn readily to the associated list of biases (3.5). This conceptualization on two levels seems to me much more convenient and heuristic than a separate verbal recital of the many possible distinct biases for each research design would be.

ET, *PT*, and *EPT* symbolize interaction terms that can be conceptualized separately [Ross and Smith 1968]. For example, *ET* denotes the possibility that experimental effects may interact with outside effects during the test period; *EPT* that experimental and pretest effects and the mere passage of time may all interact. However, these interaction terms unduly complicate the comparisons of biases; and furthermore the effects of time's passage cannot readily be separated in practice. Without sacrificing accuracy, we can reduce by three the number of explicit terms by combining the interaction terms induced by the passage of time. Thus we shall use $E_t = (E + ET)$, $P_t = (P + PT)$, and $EP_t = (EP + EPT)$. Furthermore, we suspect that P_t and EP_t are seldom important or measurable.

The expected (or mean) value for [.*Ob*] of module III is $(M + T)_b$, whereas for [.*Ex*] of module I it is $(M + E_t + T)_x$, including the experimental effect. For $[BO\tilde{b}]$ of module IV it is $(T + P_t)_b$; and for $(XE\tilde{x})$ of module II it is $(E + T + P_t + EP_t)_x$. We can expand fully this longest model as an example

$$\begin{aligned}\text{Exp}[\tilde{x} - X] &= E + T + P + ET + EP + PT + EPT \\ &= (E + ET) + T + (P + PT) + (EP + EPT) \\ &= E_t + T + P_t + (EP)_t.\end{aligned}$$

These are expected, i.e., average, effects from universes of similar individual observations on elements. For example, the effect for single pretest observations of module II can be denoted as $X_i = X + \varepsilon_i$, for an empirical value with random error. The expected or mean values of the X_i values is M_x both before and after treatment and they cancel out; the effect of time's passage on the mean M_x is taken on by T_x.

The ε_i include both individual deviations and errors of observations, representing the residual variation unaccounted for by the systematic effects of the other sources, denoted as mean biases. The ε_i are assumed to have no interaction and no covariance with those other effects. In the population (or universe) they are assumed to have mean value of zero, and a mean square value of S^2. The actual effect of this random variation on specific samples is an unknown random variable; but its expectation is zero and its squared expectation is S^2 per element in the variances for the modules and for research designs. It contributes S^2/n to the mean square error of a single mean based on n observations, as in modules I and III. To the correlated n pairs of observations of modules II and IV it contributes $(S^2 + S^2 - 2RS^2)/n = 2S^2(1 - R)/n$.

This simple error structure is brief and prevalent in the literature. With its assumptions it avoids some complexities of the real world. The error S^2/n assumes simple random samples of n elements; this can be adjusted for with design effects (7.1). Instead of a simple error term ε_i, it would be more realistic to pose error terms for each of the four component sources. However, the separate measurement and analysis of each of the components would take us into complicated statistical analyses and away from our principal goals.

3.4 FIVE BASIC DESIGNS FOR COMPARISONS

3.4A One-Shot Case Study: [.*Ex*]

"This design does not merit the title of experiment, and is introduced only as reference point" [Campbell, 1957]. The "one-shot" study refers to simple means (or medians, rates, or other descriptive statistics) based on the treatment groups E, without an explicit control group O in the design. The symbol E denotes some "experimental treatment" that precedes the measurements, whose results are observed by x. Design 1 without explicit controls receives no standing or defense in the literature. However, in practical work the means, proportions, and rates of interesting groups often serve as sources for new research ideas and further investigations and even as bases for tentative conclusions. The result x of the experimental treatment is usually compared with some standard *base* or bases computed from recognized population data, from models and projections from theory, or from subjective estimates and guesses. The comparisons may be made explicit, but they are more often merely left implicit.

Consider several examples. (1) The quality of a new batch (perhaps using new materials or processes) of a manufactured product (say electric bulbs) is compared with well-accepted standards, perhaps based on continuous process control. Such is the basis of quality control in manufacturing. (2) An intensive family planning program is introduced into a "typical" province; later the province's birth rates are compared with those of similar but "untreated" provinces. (3) Birth rates of special ethnic or religious groups or other subclasses are contrasted with accepted standard rates for comparable populations. (4) Mortality and morbidity rates of a special group are contrasted with accepted standard rates from the entire population or from comparable subclasses. In (1) and (2) the experimental treatments are deliberately introduced, and efforts can be made perhaps to randomize that introduction over the entire population, or to design some useful substitute (3.1). In (3) and (4) the E represents only observed predictor variables, and the study merely takes advantage of observed differences in the population. In both cases, if the predictor ("experimental" E) population is much larger than the study both needs and can afford, a sample design is suggested. One should try then for a probability selection of the sample within that population, rather than just a judgment or haphazard choice. In situations like (3) and (4), where the treatment is not randomized, probability selection of the sample from the population may still be feasible.

The mean square error of the difference (x-base), between the treatment mean and the base chosen as standard for comparison, may be expressed as:

$$\text{Mse}(x - \text{base}) = \frac{S^2}{n} + \text{Var}(\text{base}) + \text{Bias}^2(x - \text{base}). \quad (3.4.1)$$

The variance of the base will often be unknown and indefinite, but usually it is reliably small compared with the other two terms; otherwise the comparison would probably not be made. But if the variance is not negligible, the situation may call for a subjective estimate. The bias of the comparison is generally both more important and more difficult to estimate. Judgment must be used to choose a standard for control that is "similar" to the treatment group, except for the absence of the predictor treatment variable. Often several diverse controls may be used to increase the power of "falsifiability" (7.6). Also, the base population(s) may be adjusted for known or suspected differences in control variables (4.5).

The preceding examples illustrate the sorts of situations where this kind of comparison seems reasonable. We need good data and reasonable agreement about the base(s) used for standard(s) of comparison. Lack of good data and confidence in the validity of the base for control makes the "one-shot" study generally suspect, and altogether despised in methodological literature. Yet it is widely used in practice to search for gross differences (x-base) between some special groups and reasonably well-known standard bases for them.

Some defenders of this simple design may claim to see "no clear and absolute difference" between the problems of assessing the Bias (x-base) of this Design 1 and those of assessing the Bias ($x - b$) of Design 3 for controls without randomization. But that is a contrary absolutist view to which I do not subscribe, in light of the problems of Design 1 compared with those of Design 3 (Table 3.4.1). Good controls for Design 3 may be difficult to arrange, but for Design 1 the controls are worse, more difficult, or unavailable in many situations.

On the other hand, Var(base) may be negligible for Design 1, whereas the variance $2S^2/n$ of Design 3 is double that of S^2/n for Design 1. Furthermore, the unit cost factor $[1 + k_3]$ for Design 3 may also nearly double that of Design 1. This latter should have (in 3.4.1) the factor $G = (1 + \text{base}/cn)$, to include the cost of obtaining the data for the base relative to the cost cn of the n elements in the x treatment; but this should often be much less than one. We should also include in G a factor for the design effects of sampling, especially when samples for Design 1 may come from large clusters.

We may note two more examples (5, 6), in addition to (1) to (4) above, for situations when base data are not needed because strong theoretical models may provide adequate bases for comparisons. (5) The recovery rate, within a specified period for a new treatment, from a disease that is otherwise known to be fatal in the base, needs no control group. (6) The success rate of an intensive program to reduce illiteracy can be judged against a practical rate of nearly zero for spontaneous rates. For these situations the "external validity" problem of representation may be more important than specified controls of Design 3, which would merely double the cost, or double the variance. Example (1) above also may often be in this category.

The expected value of $[.Ex]$ in Table 3.3.1 is $E_{tx} + M_x + T_x$, where E_{tx} denotes the experimental treatment we desire to measure. M_x is the mean

TABLE 3.4.1. Biases, Costs, and Variances for Five Designs

1 Name of Design	2 Module Symbols	3 Compares	4 Bias	5 Unit Factors Variance	× Cost
1. One-shot	[.*Ex*]	*x*-base	$M_x - M_{\text{base}} + T_x - T_{\text{base}}$	1 + Var(base)	1 + base/*cn*
2. One-group, pre/post	[$XE\tilde{x}$]	$\tilde{x} - X$	$T_x + P_{tx} + PE_{tx}$	$2(1 - R)$	$1 + k_2$
3. Control group	[.*Ex*] [.*Ob*]	$x - b$	$(M_x - M_b) + (T_x - T_b)$	2	$1 + k_3$
4. Pre/post control group	[$XE\tilde{x}$] [BOb]	$(\tilde{x} - X) - (\tilde{b} - B)$	$(T_x - T_b) + (P_{tx} - P_{tb}) + PE_t$	$4(1 - R)$	$1 + k_3 + k_2 + k_4$
5. Four-group control	[$XE\tilde{x}$] [BOb] [.*Ex*] [.*Ob*]	$(\tilde{x} - X) - (\tilde{b} - B)$ or $(\tilde{x} + x/2 - X)$ $- (\tilde{b} + b/2 - B)$	$(T_x - T_b)$	$4(1 - R)$ or $1 + 2(1 - R)$	$2 + 2k_3 + k_2 + k_4$

value of the experimental group before E_{tx} occurs, and $(M_x - M_{\text{base}})$ denotes the bias due to the divergence of the experimental set from the standard base used for comparison. $(T_x - T_{\text{base}})$ denotes different effects on the two sets due to other events occurring during the period of the experiment. Thus $(M_x - M_{\text{base}}) + (T_x - T_{\text{base}})$ denotes the bias of Design 1. Furthermore, the sampling question concerning E_{tx} is especially important: To what degree is it conditional on and specific to the treatment group, and not representative of potential effects in the target population?

We should also ask when possible, as in situations such as examples 1 and 2 above, whether pretreatment observations can be introduced, so that we may have Design 2 instead of Design 1.

3.4B One-Group Pre/Post Design 2: [$XE\tilde{x}$]

Here X and $\tilde{x}$ denote the results of measurements made on the same group of elements before and after the experimental treatment E. This design consists of only module II and its mean square error may be written as:

$$\text{Mse}(\tilde{x} - X) = \frac{2(1 - R)S^2}{n}(1 + k_2) + (T + P_t + PE_t)^2. \quad (3.4.2)$$

The variance term is modified by the cost factor $(1 + k_2)$. The basic element cost is c for the experimental treatment plus the posttest plus the recruitment cost; the pretest adds ck_2 to the element cost. The total cost for n elements is $cn(1 + k_2)$. To pay for the pretest for a fixed basic cost of cn we should reduce the sample size from n to $n/(1 + k_2)$. S^2 represents the variances for both pretest and posttest observations, and R denotes the correlation coefficient between the pairs of observations on the same set of elements. As against the variance S^2/n of the basic module I, the variance, $S^2 + S^2 - 2RS^2 = 2(1 - R)S^2$, of this design differs by the factor $2(1 - R)(1 + k_2)$. Usually, $0 < k_2 < 1$ and the factor is between $2(1 - R)$ and $4(1 - R)$; hence we need between $R > 0.5$ and $R > 0.75$ to have $\text{Var}(X - \tilde{x}) < \text{Var}(x)$. High values of R are not rare when behavior is consistent and errors of observation are not high; hence Design 2 may often have smaller variance than Design 1 for the same cost. On the other hand, low values of R are also common, especially in attitudinal and psychological observations, because of the instability of variables and errors in observations; then perhaps $2(1 - R)(1 + k_2) > 1$ and Design 3 may have higher variance than Design 1.

In small samples the variance term may predominate. But the bias term becomes more important in larger samples, and more difficult to guess—both before and after the observations. The design has the expectation $(E_t + T + P_t + PE_t)_x$; and the two components $P_{tx} + PE_{tx}$ arising from the pretest can often be supposed to be small compared with the other two. Since the same elements were used for both observations, the component for the group means cancels out: $(M_x - M_x) = 0$. However, the two components $E_{tx} + T_x$ may both be large and it may be difficult to disentangle the time effects T_x from the treatment effects E_{tx}. If time is short and supposed to be without catastrophic events, we may consider T_x to be small compared with E_{tx}. Without such assurances one may resort to Design 3 or to Design 4, both with controls of T_b for time effects in the comparisons. However, these two designs, especially Design 4, increase variances and costs also.

Therefore one may try to obtain control of time effects for T from some standard base, as for Design 1, if these data are available. Furthermore, the question of representation may also become important and the subscripts x are used to call attention to it. If the group is not selected with probability sampling from the target population, inference from E_{tx} and T_x to the effect of T in the population becomes subject to judgment about their representativeness.

An interesting extension concerns possible situations where the posttest is only a sample of the pretest (or vice versa). Let n_l and n_s denote respectively the larger and smaller samples; then the variance of the comparison [Kish 1965a, 12.4] is:

$$\mathrm{Var}(\tilde{x} - X) = S^2\left[\frac{1}{n_s} + \frac{1 - 2R}{n_l}\right].$$

When $R > 0.5$, the larger sample reduces the second term in the variance. But this term is positive when $R < 0.5$, and the excess $(n_l - n_s)$ cases would actually increase the variance; it may be better to retain only the common portion n_s with the variance $2(1 - R)S^2/n_s$, and discard the surplus $(n_l - n_s)$ observations, though this may seem counterintuitive. But there are estimators other than $(\tilde{x} - X)$ that can utilize excess observations in both pretest and posttest.

The preceding amounts to two new designs: either $[XE\tilde{x}]_s + [BO.]_{1-s}$ or $[XE\tilde{x}]_s + [.Ob]_{1-s}$. Another possibility would be $[.Ex] + [BO.]$, with the pre- and postobservations made on "similar" but separate samples; the correlation R would be absent from this design for two separated samples. These are illustrations of the possible extensions of the basic methodological approach to which the symbols may be extended and modules constructed.

3.4C Control Group Comparison Design 3: [.*Ex*] + [.*Ob*]

When people begin to think of experiments, they first imagine something like this design, which combines module I plus III: an experimental group, exposed to treatment E, yields results x, whereas a control group, exposed to a control treatment, or to "nothing," yields results b. The virtues of this design depend greatly on whether or not the treatment and control groups were randomized or how well randomization was approximated. *Internal validity* demands that both samples be representative of the *same* population. *External validity* requires that both samples be representative of the *same target* population. Randomizations in two senses are the keys to achieving each kind of validity (1.4). Randomized separation of treatment and control groups is a critical distinction, which separates "true" ("ideal") experiments from other investigations ("quasi-experiments"). Campbell [1957, 1963] denotes this distinction with separate names: (1) *posttest only control* versus (2) *static-group* comparison. Here we shall distinguish randomized Design 3 with the symbols [.*Ex*] + [.*Øc*].

The expected difference in Design 3 is

$$\begin{aligned}\text{Exp}(x - b) &= [(E_t + M + T)_x - (M + T)_b] \\ &= E_{tx} + [(M_x - M_b) + (T_x - T_b)].\end{aligned}$$

Thus the mean square error of Design 3 may be denoted with

$$\text{MSE}(x - b) = \frac{2S^2}{n}[1 + k_3] + [(M_x - M_b) + (T_x - T_b)]^2. \quad (3.4.3)$$

For the variance term here we assumed two independent simple random samples of n from populations with element variances S^2. The cost for the two samples of size n is $cn + ck_3n = cn(1 + k_3)$. The element cost ck_3 for the control may be somewhat less than for the experimental group: then $0 < k_3 < 1$. For a comparable fixed cost cn, the sample size n must be reduced by the factor $2(1 + k_3)$; then the variance of Design 3 is $2(1 + k_3)$ times greater than for the basic Design 1. For Design 2 over Design 3 the variances are in the ratio $(1 - R)(1 + k_2)/(1 + k_3)$. This can be much less than 1, especially when R is large and $k_2 < k_3$. Then Design 2 should be preferred over Design 3 on these considerations of variance alone.

The variance of Design 3 may be reduced somewhat by avoiding complete independence of the two samples with some type of "matching" of the two samples. If we denote the element variance reduced with matching as $(1 - r_3)S^2$, the ratio of Design 2 over Design 3 becomes $(1 - R)(1 + k_2)/(1 - r_3)(1 + k_3)$, likely still less than 1 and still in favor of Design 2. Usually the gains in the variance are smaller than hoped for; but matching may also reduce the bias of nonrandomized selections and that may be more important.

The bias of Design 3, $(M_x - M_b) + (T_x - T_b)$, *may* often be considerably greater than the bias $T_x + P_t + PE_t$ in Design 2. The effects of pretests may be negligible; and T_x as well as $(T_x - T_b)$ may be unimportant, though here the advantage may be with Design 3. However, efforts to obtain good controls to reduce $(M_x - M_b)$ with confidence may be very difficult.

In "ideal" experiments we would have $(M_x - M_b) = 0$ and $(T_x - T_b) = 0$: Complete randomization should give us the same expectations for both treatment and control groups in Design 3. Only selection biases may remain in E_{tx} to the degree that the groups were not representative, because not randomized over the target population. However, we must remember that for a randomized experiment it is difficult to achieve probability sampling.

If we would disregard the problem of external validity, randomized Design 3 seems preferable to Design 2, which has bias T_x, plus conceivably $P_t + PE_t$. We may see that preference expressed for this design by Campbell [1957, 1963], Ross and Smith [1968], and Namboodiri [1970]. But this reduced bias should be balanced against the reduction in the variance by $(1 - R)(1 + k_2)/(1 + k_3)$ for Design 2. Furthermore, probability sampling may be easier to achieve for Design 2 than for Design 3.

In practice, improved controls, preferably randomization, face two obstacles in social research—and in many other kinds as well. Cost is one. The other is the dilemma between the goal of reducing the bias, especially $(M_x - M_b)$ for better internal validity, and the goal of greater external validity by spreading the sample over the target population, preferably with randomization. I do not accept the claims of clear and universal priority for internal versus external validity (3.5).

In the face of these conflicts perhaps we can introduce a combination of two studies. First, a study with Design 2, preferably randomized over the population; then another study with Design 3, perhaps on a smaller scale. But this must be compared with the costs and expected results of Design 4.

3.4D Pre/Post Control Group Design 4: $[XE\tilde{x}] + [BO\tilde{b}]$

This more elaborate design, combining moduls II and IV, has been introduced to overcome both kinds of biases present in Designs 2 and 3. "For these reasons, the Pretest-Post-test Control Group Design has been the ideal in the social sciences for some thirty years" [Campbell, 1957]. Like other ideals, I suspect it is more widely urged in the literature than actually practiced in the field. Like other ideals, it also demands sacrifices: Variances and costs are both subject to increases by factors of about 2 to 4, hence the unit variance × cost factor increases by about 4 to 16—unless R is large and $k_2 + k_4$ is small.

The mean square error of Design 4 may be denoted with

$$\text{Mse}\,[(\tilde{x} - X) - (\tilde{b} - B)] = 4(1 - R)\frac{S^2}{n}[1 + k_2 + k_3 + k_4] + [(T_x - T_b) + (P_{tx} - P_{tb}) + PE_{tx}]^2. \tag{3.4.4}$$

We assumed independence between the two modules and simple random samples of n from the population, with element variances of S^2 for both. The cost of the design is $cn(1 + k_2 + k_3 + k_4)$. When this is equated to the cost fixed at cn for the basic design 1, we see that $n_4 = n/(1 + k_2 + k_3 + k_4)$, and the variance is correspondingly increased. Often $k_3 < 1$, and $k_2 = k_4 < 1$, perhaps close to zero. The variance is increased over Design 1 by some factor between $4(1 - R)$ and $16(1 - R)$, and perhaps $8(1 - R)$ may be closer than either extreme; and the increase depends on how high R is.

The variance for Design 4 has the factor $2(1 + k_2 + k_3 + k_4)/(1 + k_2)$ over Design 2. Thus one may be paying with a variance about four times greater for differencing the sources of bias $(T_x + P_t)$ in Design 2. Compared to Design 3 the variance of Design 4 has the factor $2(1 - R)(1 + k_2 + k_3 + k_4)/(1 + k_3)$; if $(k_2 + k_4)$ and $(1 - R)$ are small, the variance for Design 4 may be less than for Design 3. Lastly, both of these ratios may be affected by reductions of the element variances induced with matched sampling; its effect on S^2 in Design 3 may not be negligible.

We are most interested in the bias term of Design 4, (in 3.4.4). This comes from the expected difference

$$\begin{aligned}\text{Exp}[(\tilde{x} - X) - (\tilde{b} - B)] &= (E_t + T + P_t + PE_t)_x - (T + P_t)_b \\ &= E_{tx} + [(T_x - T_b) + (P_{tx} - P_{tb}) + PE_{tx}].\end{aligned}$$

In the "ideal" experiment, symbolized with $[XE\tilde{x}] + [CØ\tilde{c}]$, randomization should remove $(T_x - T_b)$ and $(P_{tx} - P_{tb})$, and should leave only PE_{tx} as (negligible) bias in Design 4. Thus the bias terms T_x and P_{tx} are removed by the addition of module IV in Design 4 to the module II of Design 2. However, Design 3 does not even have the marginal bias term PE_{tx}, from interactions of the pretest P with E and T. This consideration, confined to randomized experiments, and the simple counting of five components of the bias term have made Design 4 disliked and Design 3 prefered by some writers [Ross and Smith 1968, Namboodiri 1970]. But I do not subscribe to that view, and consider that bias components and designs should be judged in the context of specific situations, and PE_{tx} usually disregarded.

We should be most interested in comparing bias terms for non-randomized designs. Here Design 4 has the second-order terms $(T_x - T_b)(P_{tx} - P_{tb})$ against the first-order terms $T_x + P_{tx}$ of Design 2. The decrease in potential bias may compensate for the increase in variance in Design 4. The bias in Design 3 is $(M_x - M_b) + (T_x - T_b)$; and $(M_x - M_b)$ may be considerably more risky than $(P_{tx} - P_{tb}) + PE_{tx}$ from the pretest effects in Design 4. Thus when there are grave doubts about lack of randomization, Design 4 may have considerable advantages over both Designs 2 and 3.

3.4E The Four-Group Design 5: $[XE\tilde{x}] + [BO\tilde{b}] + [.Ex] + [.Ob]$

Design 5 uses all four modules: it adds the pair I and III for Design 3 to the pair II and IV of Design 4. "This Solomon Four-Group Design enables one both to control and measure both the main and interaction effects of testing and the main effects of a composite of maturation and history. It has become the new ideal for social scientists" [Campbell 1957]. The design aims at separating and measuring the effects of testing, present in Design 4 but absent in Design 3, from other effects due to time's passage. It may be adapted when the effects of testing are suspected of becoming potentially important. It gets strong support [Campbell 1957; Campbell and Stanley 1963; Ross and Smith 1968] for its emphasis on validity and on cautions against possible biases from the effects of pretests.

The design originated in the literature for educational experiments; there, the effects P of pretesting may sometimes conceivably become large enough to be considered, as compared with experimental results E or other sources of bias. There also, the randomization of the four modules may be feasible. But it is seldom used in actual research because of its high cost, its high variance, and its complexity, and because testing effects are seldom that important compared with other possible sources of bias and errors. This is another "ideal" out-of-work.

Furthermore: "There is no singular statistical procedure which makes use of all six sets of observations simultaneously. The asymmetrics of the design rule out the analysis of variance of gain scores" [Campbell and Stanley 1963]. Moreover, following the fashion of the times, discussions of statistical procedures are confined to testing of null hypotheses of zero differences. One procedure [Campbell and Stanley 1963] calls for testing separately the modules $[XE\tilde{x}] - [BO\tilde{b}] = [(\tilde{x} - X) - (\tilde{b} - B)]$, which has, as in

Design 4, expectation $E_t + PE_t$, with $(T_x - T_b) = 0$ and $(P_{tx} - P_{tb}) = 0$ in randomized experiments. Then from the posttest, measurements of all four modules $[(\tilde{x} - x) - (\tilde{b} - b)]$ may be computed, from which in randomized experiments PE_t may be obtained to "measure both the main and interaction effects of testing . . .," the chief aim of this design. That aim is not often foremost in research, particularly in nonrandomized quasi-experiments, when other sources of bias seem more important. Different comparisons yield diverse components, and, unfortunately, there appears to be no uniformly best way to test for them and to utilize the results of this design.

Perhaps Design 5 should be viewed as an effort to use Designs 4 and 3 in combination, when neither seems clearly the best. As a compromise between the two designs, we may take both, though not necessarily in equal proportions, especially if the bias threats and the costs of the two designs differ greatly. The comparisons for the two designs may be contrasted, searched for clues, and, with luck, courage, and caution, combined into a joint result.

Different combinations may also be suggested, such as Designs 2 and 3 when neither is clearly better. Still others may also be tried, given the flexibility of the modular method. At the end of 3.4B we also noted some possible modifications of Design 2, also built from the modules. In 3.4A the possible uses of standard bases was introduced. In 3.6 we shall note extensions of the observations over several and longer time periods. The approach, the tools, and the method can be usefully adapted to other uses.

3.5 CLASSIFICATION FOR 22 SOURCES OF BIAS

A complete, theoretically sound, logical taxonomy of all biases, good for all specific situations in any field of research cannot be achieved. The list in Table 3.5.1 of 22 sources of bias is aimed at providing reminders, thus helping the researchers themselves find and describe their actual sources of potential biases in specific situations. These sources thus serve as bridges from the major types of errors, which distinguish research designs, to the many specific variations that may occur in actual situations. These 22 sources at level III (3.3B) bridge the gap between the four types at abstract level II and the specific biases of levels IV and V, which are too numerous to name and describe. With this two-level framwork I hope to introduce comprehensibility into what otherwise would appear a chaotic mess of too many different sorts of biases. Most of us can grasp four or six concepts at one time, but 22 seems too many.

These 22 sources denote a personal attempt at classification, with arbitrary names and descriptions. Readers may change them, name them,

TABLE 3.5.1. Classification for 22 Sources of Bias in Six Major Types[a]

T	*Time effects*
*T*1	History [1] of external events, common, unique, unforeseen
*T*2	Maturation [2] of subjects, internal, individual, gradual, aging, tiring, learning
*T*3	Instrumentation [4]: changes in measurements, standards, observers
*T*4	Treatment changes over time in panel studies
M	*Selection of treatment members*
*M*1	Differential selection [6] of members for groups
*M*2	Differential loss [7]: due to mortality, nonresponse, migration
*M*3	Selection–maturation interaction [8]
*M*4	Regression effects [5] from selection of extremes for comparisons
P	*Pretesting*
*P*1	Pretest effects [3]: denoted by *P* in text
*P*2	Pretest–experimental interaction [9]: denoted by *PE* and *PET*
*P*3	Pretest–time interaction: denoted by *PT*
E	*Experimental treatments: predictor variables*
*E*1	Artificiality of experimental treatments [11] and environment
*E*2	Multitreatment interference [12]
*E*3	Errors in measuring treatments [4]
R	*Errors in measured responses: predictand variables*
*R*1	Artificial responses [11] instead of realism
*R*2	Timing of posttests
*R*3	Neglected responses: side effects, harmful or beneficial
*R*4	Differences in posttest measures between treatment groups [4]
U	*Representation of target population (universe)*
*U*1	Divergence of coverage [10] from frame to target population
*U*2	Losses due to mortality, nonresponse, migration
*U*3	Change of target population over time
*U*4	Environmental limitations

[a]Numbers in brackets refer to "12 threats to validity" in Campbell [1957, 1963].

and construct their own. However, some consistency is desirable, and I followed the best-known list of "12 factors jeopardizing the validity of various experimental designs" [Campbell and Stanley 1963], based on an earlier effort [Campbell 1957]. Those 12 factors are shown with their numbers in brackets in Table 3.5.1 and are described at the end of this section. The other 10 represent additions that I felt were needed. The division into six major types is mine, including the four used in Section 3.3 plus two new ones.

The effects of time's passage, type *T*, can have many forms and I divided them into four sources. Source $T1$ refers to sources that are *external* to the subjects and that also may have common effects on all subjects and in both (or all) treatment groups. Source $T2$ refers to changes *internal* to the subjects, which may be as diverse as learning or tiring or merely aging. This internal source $T2$ is more likely to be gradual, whereas the external effects $T1$ may be sudden, unforeseen, and unique. But these features may be reversed and I list them only to alert readers and researchers to them. Type $T3$ refers to changes in *measurements* that may be introduced deliberately, but more often just creep in despite attempts to keep standards stable and unchanged. Type IV, on the other hand, refers to changes in the *treatments* themselves: changes that may occur in repeated treatments, even when unintended. Furthermore, there may be interactions among these four sources. For example, the "same" treatment may appear different in changed external circumstances or to tired, aged, or more learned subjects.

All events take place in "the river of time's flow," and some treatments may need considerable spans of time, which may also permit many other extraneous effects to develop. Furthermore, the effects of time can produce interactions with other sources of bias; some of these interactions appear below as separate sources of bias. For example, the bias $M3$ of selection–maturation interaction [8] refers to the possibility that learning or tiring may be greater (or less) in the treatment group than in the control group. Losses from samples, $M2$, may also be different in the two groups. These differences can have biasing effects on the means when they are selective in the response variable.

Type M biases are due to differences in the selection of individuals between the treatment and the control groups; in general, between the several treatment groups. Differential selection $M1$ occurs before treatments begin; this poses the gravest problems for nonrandomized observational studies, in contrast to the equality introduced by randomization in true experiments. On the other hand, differential loss $M2$ is a potential bias that can occur in both kinds of designs, when differences in treatments result in differential out-selection due to nonresponse, migration, or mortality. Selection–maturation interaction $M3$ may occur in nonrandomized observational studies when maturation (learning, tiring) occurs at different rates in the groups that were "matched" artificially at the beginning of the study. "Regression effects" $M4$ may occur when groups have been selected on the basis of extreme scores, and those scores hide random components that are subject to "regression toward the mean" on remeasurement regardless of treatment. This serves as an example when disturbing factors (1.2) from a nonrandomized selection process contribute to a potential bias that is confounded with the explanatory factors. Confusing selection procedures with the explanatory factors is the common feature of these diverse sources of bias within this major type M; they all interfere with the "internal validity" of non-randomized studies.

Potential bias sources of type P may arise from pretesting observations before the treatment, as in modules II and IV, but not in I and III. However, pretest observations can often reduce variances drastically and for relatively low cost (3.3). Thus we may face a conflict if considerable effects seem possible from the pretest observations $P1$ and from their interactions with other sources of bias. Here we distinguish the major pretest effects $P1$, denoted as P in Section 3.3, the pretest–experimental interactions, denoted there by PE and PET; and the pretest–time interactions, denoted by PT. Here we follow the literature of biases with this elaborate structure [Campbell 1957; Campbell and Stanley 1963; Ross and Smith 1968; Namboodiri 1970]. However, I expect that in most situations these sources will tend to be negligible compared with other considerable sources of bias and variance. In cases where fear of considerable P effects seem reasonable I suggest an innovation (untried to my knowledge): Make the pretest as innocuous as possible so long as that minimal pretest can yield the desired baseline for the correlations R with the posttests, because this is its main or only function.

Experimental treatments may have unplanned, and even unexpected, side effects. This may be true for predictor variables in general, whether they are considered as treatments, or stimuli, or causal variables. The artificiality of experimental treatments $E1$ and of their environmental setting poses vast problems, for which we can offer no general advice that would also be helpful in specific settings. It is worthwhile to call attention to $E2$ [12] "multiple-treatment interference, likely to occur whenever multiple treatments are applied to the same respondents, because the effects of prior treatments are not usually erasable." Errors in measuring instruments $E3$ or "instrumentation" [4] is included for a general caution, because there are many examples of unexpected changes in responses from unintended, slight changes in meaning, in questionnaires, in perceptions, and generally in stimuli. These cautions overlap with others such as $M3$ and $T3$. In general, changes of types E and T cannot be distinguished, and we proposed (in 3.3B) to consider $E_t = E + ET$ jointly.

As with predictors, potential instrumental biases exist when measuring in posttests the response variables R, i.e., the predictand, criteria, outcome, effect variables. Artificial responses, $R1$, encompass in two words probably the most daunting problems facing all "invasive" research—and in experiments often probably even more than in controlled observations. The criterion of realism in measuring effects is probably a principal reason for the frequent use of observational studies instead of experiments in social and medical research. We must admit that with this type of bias we are beyond the narrow confines of statistical design and are mostly within the substantive domains of specific disciplines. However, the lines of responsibility should not and cannot be drawn sharply, and statistical design and "concept validity" must interact in practice. Indeed, this book began (1.1) with this obstacle to true experiments, which—in addition to cost factors and ethical considerations—lead so often to the use of controlled observations.

Two kinds of biases in response measurements merit special attention. First, the timing of the posttests, $R2$, often has drastic effects on measures of the consequences of treatments. Short-term success rates may not only be quantitatively different from long-term rates; they may also be qualitatively different in direction. This problem receives separate and longer attention later (3.6). Second, neglected responses $R3$ introduce related and even broader questions; they are substantive questions but they have been also thrown at (the feet or heads of) statisticians: What good is measuring precisely and validly one specified effect and letting other, perhaps more important, effects go unmeasured and neglected? Furthermore, it is true that statistical multivariate analysis often uses multivariate predictors and only a single predictand (dependent) variable. From the practical side of medical and social research we also hear of research results for main effects, which neglect "side effects." (See Figure 3.6.2B) Furthermore, with the label "side effects" people usually refer only to harmful effects, neglecting possible beneficial side effects. The history of contraceptive pills provides one example where harmful side effects may well be outweighed by beneficial side effects [Weller and Bouvier 1981, Ch. 6], in terms of not only medical health but also great social benefits on a world scale.

Differences between treatment and control groups in actual posttest measures $R4$ are well known in the field of clinical trials. "Double-blind clinical trials" are prescribed as standard protection against unintended biases in diagnoses, usually in favor of new medicines, new treatments, and new operations [Gilbert, Light, and Mosteller 1975; Kahn 1972; Meier 1972]. This is an old problem in fields of medicine; and we suspect that it is no less important in social research, where there is as much room for false optimism and perhaps even more scope for self-delusion (both for subjects and for researchers) than in the effects of pills, vaccines, and operations. I guess that the only reason for fewer instances (if any?) of "double-blind" trials in social research is that the nature of social treatments is much more difficult (or impossible?) to hide from subjects and from researchers than the identity of pills and injections.

Problems of representation (U) are commonly ignored in the literature of both experimental design and controlled observations; sometimes they are even denied. However, we can be brief here, because they were treated extensively in Sections 1.7, 2.7 and 3.1. Divergence of coverage [10] from the frame of selection to the target population *U*1 has been noted in the context of representation (2.1). But after selection, further divergence also occurs, often wide, because of losses from nonresponses, migration, and mortality *U*2, particularly in panel studies. Both *U*1 and *U*2 biases occur in the sample, but *U*3 points to changes in the target population. Following a "moving-target" population over a time span raises questions of inference to populations beyond the target population (2.1, 6.3). This also applies to environmental limitations *U*4, which confine any population the researcher is able to cover in the target population.

Readers can profit here from a direct look at the source and the list on which mine was based [Campbell and Stanley 1963].

> Fundamental to this listing is a distinction between internal validity and external validity. *Internal validity* is the basic minimum without which any experiment is uninterpretable: Did in fact the experimental treatments make a difference in this specific experimental instance? *External validity* asks the question of *generalizability*: To what populations, settings, treatment variables, and measurement variables can this effect be generalized? Both types of criteria are obviously important, even though they are frequently at odds in that features increasing one may jeopardize the other. While internal validity is the *sine qua non*, and while the question of external validity, like the question of inductive inference, is never completely answerable, the selection of designs strong in both types of validity is obviously our ideal.
>
> Relevant to *internal validity*, eight different classes of extraneous variables will be presented; these variables, if not controlled in the experimental design, might produce effects confounded with the effect of the experimental stimulus. They represent the effects of:
>
> 1. *History*, the specific events occurring between the first and second measurement in addition to the experimental variable.
> 2. *Maturation*, processes within the respondents operating as a function of the passage of time per se (not specific to the particular events), including growing older, growing hungrier, growing more tired, and the like.
> 3. *Testing*, the effects of taking a test upon the scores of a second testing.
> 4. *Instrumentation*, in which changes in the calibration of a measuring instrument or changes in the observers or scorers used may produce changes in the obtained measurements.
> 5. *Statistical regression*, operating where groups have been selected on the basis of their extreme scores.
> 6. Biases resulting in differential *selection* of respondents for the comparison groups.
> 7. *Experimental mortality*, or differential loss of respondents from the comparison groups.

8. *Selection–maturation interaction*, etc., which in certain of the multiple-group quasi-experimental designs, is confounded with, i.e., might be mistaken for, the effect of the experimental variable.

The factors jeopardizing *external validity* or *representativeness* which will be discussed are:

9. The reactive or *interaction effect* of testing, in which a pretest might increase or decrease the respondent's sensitivity or responsiveness to the experimental variable and thus make the results obtained for a pretested population unrepresentative of the effects of the experimental variable for the unpretested universe from which the experimental respondents were selected.
10. The *interaction effects* of *selection* biases and the *experimental variable*.
11. *Reactive effects* of *experimental arrangements*, which would preclude generalization about the effect of the experimental variable upon persons being exposed to it in nonexperimental settings.
12. *Multiple-treatment interference*, likely to occur whenever multiple treatments are applied to the same respondents, because the effects of prior treatments are not usually erasable.

A modified and longer list and its source [Cook and Campbell 1976] are also worth studying. The list is placed in a structure of

> four kinds of validity. Statistical conclusion validity refers to the validity of conclusions we draw on the basis of statistical evidence about whether a presumed cause and effect co-vary; internal validity refers to the validity of any conclusions we draw about whether a demonstrated statistical relationship implies cause; construct validity refers to the validity with which cause the effect operations are labeled in theory-relevant or generalizable terms; and external validity refers to the validity with which a causal relationship can be generalized across persons, settings, and times.

Both sources [also Campbell 1957] give good expositions of "threats to validity" linked to specific basic designs, together with many empirical examples and long lists of references.

3.6 TIME CURVES OF RESPONSES

> We shall conceive of the causal relation in the only way in which we are able to make sense of it: *operationally*. In other words, the relation between cause and effect is regarded like the relation between turning the steering wheel in a car and the turning of the wheels themselves....
>
> By means of experiment one hopes to find relations between independent and dependent variables, and these relations are then often referred to

TABLE 3.6.1. Live Births per 1000 Inhabitants, Romania, 1960, 1965–70[a]

Quarters	1960	1965	1966	1967	1968	1969	1970
I	20.0	15.3	14.7	15.9	29.6	25.1	21.5
II	19.9	15.6	15.0	22.8	27.0	23.9	22.7
III	19.5	14.7	14.4	39.1	26.6	24.0	21.2
IV	17.4	13.1	13.3	31.7	24.0	20.6	19.2
Annual rate	19.1	14.6	14.3	27.4	26.7	23.3	21.1

[a]From Teitelbaum, 1972.

> as causal under the following four conditions: (1) that they are immediate, i.e., that there is no appreciable time-lag between changes on the independent and the dependent sides, (2) that they are deterministic, . . . (3) monotone, . . . (4) invariant. . . . [Galtung 1975]

In statistical design for social research we should plan for typical situations that have *none of the above* characteristics. Specifically, in this section we discuss designs for dealing with "appreciable time-lags," because we must measure effects that are far from "immediate" after the application of treatments. Diverse lags and successive changes of effects occur commonly, not only in social research but also in the medical and biological sciences. Even chemical reactions take time and even the speed of light is finite, but they are not our present concern. Here we discuss the timing of the posttreatment observations for the modules and designs of 3.3 and 3.4.

We may begin with an instructive and colorful example: the fertility effects in Romania, after legal abortions were dramatically curtailed on 1 November 1966, are shown in Table 3.6.1 and Figure 3.6.2a [Teitelbaum 1972]. We note a sharp rise, almost a tripling, of the birthrates after the natural gap of nine months for the population surprised by the new law. But in three years the population responded gradually with illegal abortions and with birthrates that erased two-thirds of the sudden rise. There is a lesson in that curve for those who would abolish legal abortion without reckoning on illegal abortions, which have been universally practiced where legal abortions are unavailable. Furthermore, Berelson [1979] showed that for 10 years (Figure 3.6.2b) Romania did get higher birthrates and higher total population, as intended, with their decreased legal abortions; although they did keep on rising slowly through the decade. But note also the dramatic rise in abortion-related deaths (due to illegal abortions); these side-effects were unplanned, but they should not have been totally unanticipated. In the longer range the "treatment" was even less effective: The crude birthrate

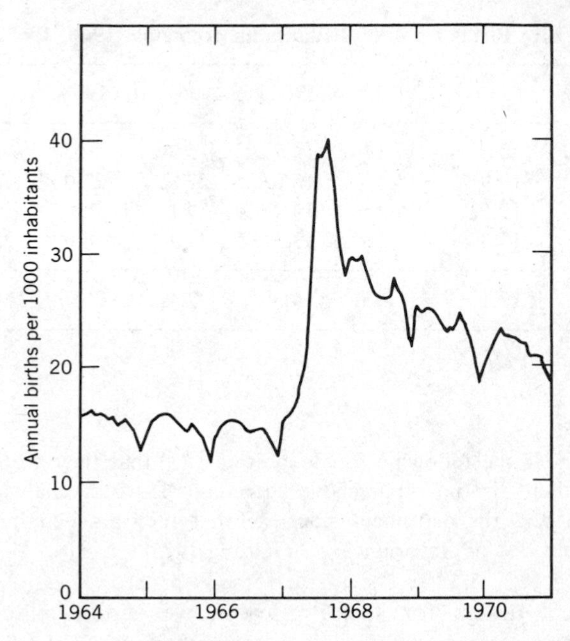

Figure 3.6.2a. Live births per thousand inhabitants: Romania, 1964–70 [Teitlebaum 1972].

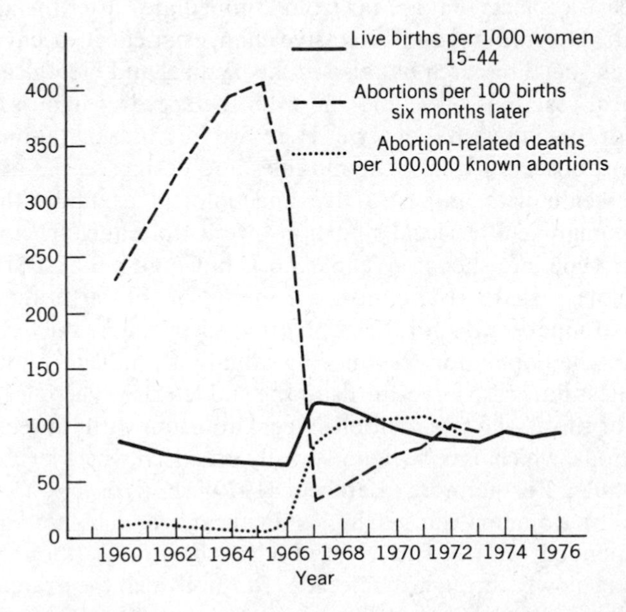

Figure 3.6.2b. Romania's 1966 anti-abortion decree: the demographic experience of the first decade [Berelson 1979].

(births per 1000 inhabitants) hovered between 18 and 20 during 1971 to 1976, but fell to 15.2 by 1982, 16 years after the abortion law of 1966.

Another interesting example arose when Sweden changed in 1967 from driving on the left side of roads and streets to the right side. Most of us feared an initial increase in traffic accidents until the confused Swedes adjusted to their new rules; and thus we were surprised by a sharp drop instead in the first month. However, this was not a permanent benefit for changing to a naturally better side of the road, but merely the unanticipated excess benefit of caution over habit. As the new habit overcame caution within a few months, traffic accidents rose back to their previous levels.

The Hawthorne effect is famous in social psychology, though its exact cause is still controversial [Kahn 1975].

> When Rothleisberger and Dickson (1939) carried out their experiments to find conditions that would maximize productivity of factory teams at the Hawthorne Works of Western Electric, they found that every change—increasing lighting or reducing it, increasing the wage scale or reducing it—seemed to increase the group productivity. Paying attention to people, which occurs in placing them in an experiment, changes their behavior. This rather unpredictable change is called the Hawthorne effect. [Mosteller 1967]

Here we note only that such effects are bound to be temporal and longer measurements are needed to separate shorter from longer effects. For example, we doubt that changing the driving lanes in Sweden every few months would keep on lowering the accident rates.

Many less dramatic examples testify to the pervasiveness of problems to which single measurements before and after treatments fail to yield satisfactory solutions.

> The essence of the time-series design is the presence of a periodic measurement process on some group or individual and the introduction of an experimental change into this time series of measurements, the results of which are indicated by a discontinuity in the measurements recorded in the time series. [Campbell and Stanley 1963, Section 7; see also Campbell 1969; Cook and Campbell 1979, Ch. 5]

The series of observations may be represented by $O_1\ O_2\ O_3\ O_4\ O_5\ O_6\ \ldots$ etc., and where to introduce the treatment X into this series is a prime problem for design. The symbols X for treatments and O for observations are consistent with the cited literature, though not with the symbols of (3.3), where they are E or O for treatments and x or b for observations. We omit the problems of analysis to separate measured effects from other possible causes of change and both from "background noise." (See Figure 3.6.3)

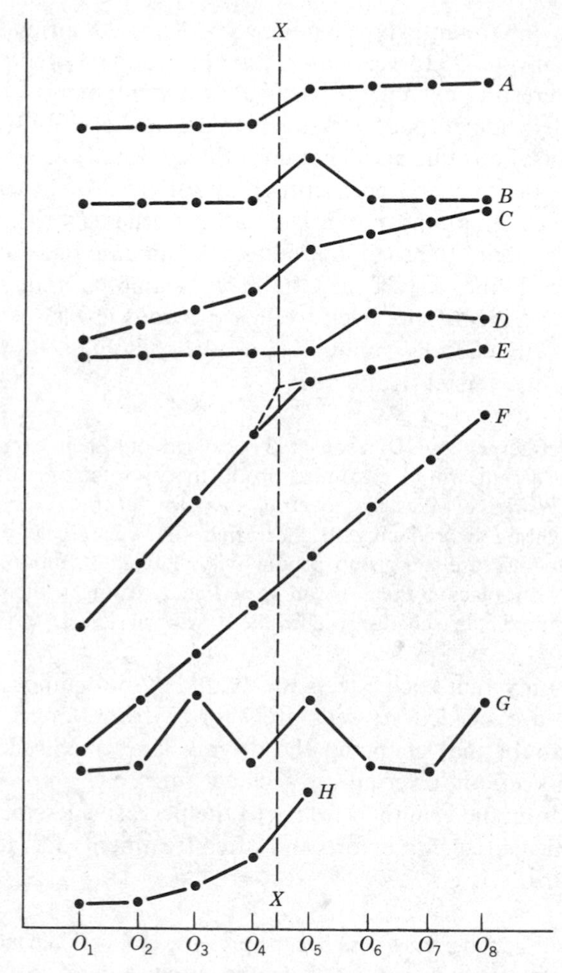

Figure 3.6.3. Some possible outcome patterns from the introduction of an experimental variable at point *X* into a time series of measurements, 0_1—0_3. Except for *D*, the 0_4—0_5 gain is the same for all time series, while the legitimacy of inferring an effect varies widely, being strongest in *A* and *B*, and totally unjustified in *F*, *G*, and *H* [Campbell and Stanley 1963].

In addition to offering general cautions about the diversity of possible outcomes, I introduce an oversimplified skeletal scheme as a framework for that diversity (Figure 3.6.4). Time of treatment is represented by *X* followed by a gap *W* for waiting times of undefined duration, which can also be compressed or expanded like an accordion.

In column 1 we see representations of the four basic types. Type *A* with

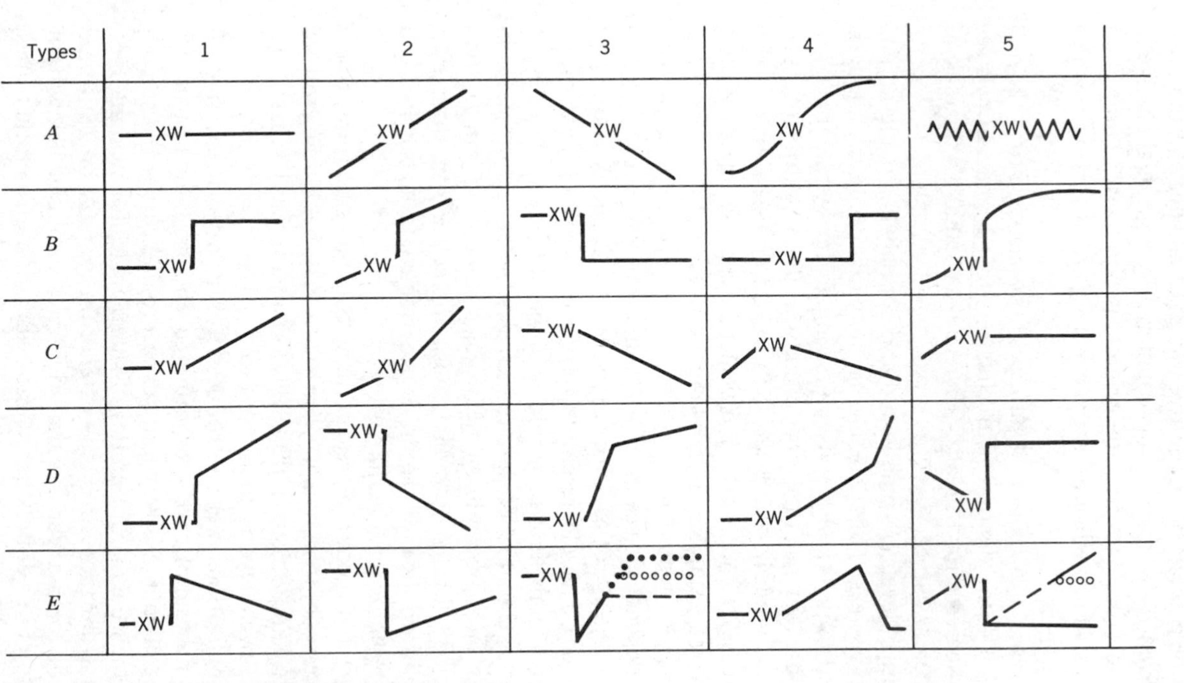

Figure 3.6.4. Types of possible changes after treatments.

These 25 cells should represent schematically most types of possible changes after an "experimental" (*X*) treatment and periods of waiting (*W*) for effects. Five major types (*A–E*) are defined in column 1, and columns 2–5 are variations on those types. The comparison between A1 and B1 is the simplest type that would be adequately measured with the before-after tests on treatment (B)/control (A) differences ($\bar{y}_b - \bar{y}_a$). The others serve as cautions to alert researchers to other types of changes which are possible. With more and longer measurements further changes can be perceivable.

no effect is contrasted with type *B* with a sharp one-time additive effect. This would be the simplest contrast for two treatments, *A* and *B*. Type *C* shows a gradual effect, and type *D* has both additive and gradual effects in the same direction. However, in *E* the sharp rise is countered by a gradual change in the opposite direction, back toward the original position. Remember that Romanian abortions and Swedish motor accidents were of this type, which may be fairly common.

For each of the types *A* to *E* I sketch subtypes for varieties. For *A* the null effect may take any form denoting no change. In *A*2 and *A*3 we see undisturbed linear increase or decrease, whereas in *A*4 and *A*5 the lack of effect occurs in different curves.

A sudden increase of levels (*B*) may be induced not only in a flat level (*B*1) but also in a steady increase (*B*2). The change may involve a drop in levels (*B*3) or a postponed rise (*B*4); also a rise in a complex (growth) curve (*B*5).

For type *C* we note that a change of slope may heighten existing increases (*C*2), but it may also result in decreases of slope (*C*3 and *C*4). It may also flatten existing increases (*C*5); for example, a successful program (*X*) of birth (or crime) control may stabilize a growing population (or crime rate). Finding actual examples would not be difficult for any of these models. All these models depend on scales of measurement and are subject to transformations of those scales. For example, an induced change of population totals like *C*1 would resemble *B*1 if measured in growth rates. Similarly, population totals like *A*2 would resemble *A*1 in growth rates.

Changes in both level and slope denote increases in *D*1 and decreases in *D*2. The changes of level are represented by steep slopes (early in *D*3 and later in *D*4), which may be difficult to distinguish in practice from sharp rises of level. In *D*5 downward slide (of profits, or achievement tests) is halted and then stabilized (at a higher level perhaps).

I suspect (or fear) that types *E* represent many common situations, where early apparent successes of treatment (*X*) are followed by backsliding toward the original levels (*E*4 and *E*5), as in the earlier examples from Romania and Sweden, or perhaps even beyond them.

Five crude simplifications were introduced for brevity. First, smooth curves replace the fluctuations we are bound to find, as hinted at in *A*5. Second, under those fluctuations we should expect to find curves and smooth transitions, rather than the discontinuities of sharp angles for straight lines. Third, we ignore how far these lines may extend; for example in *A*5, would the descent flatten out before reaching the origin at *X* or plunge below it, or return to it? And so on, to doubts about the destination of the other lines. Fourth, the figures show only simple additive jumps or changes in the slope of linear changes; instead, the main interest may be in acceleration or deceleration, i.e., in changes in the rates of change. For example, can inflation (change of prices) or population growth be slowed down? And so on, to higher orders of change. Fifth, we discuss the effects of treatments at a single point of time, whereas repeated or continuing treatments may have different and complex patterns of response.

A complicated and dramatic example of reversals of subtypes *E*5 concerns a treatment for Parkinsonism and for encephalitis lethargica (sleeping sickness), "L-DOPA is a ... true miracle drug of our age" was said and believed for a while after its first uses in 1967. Years later its results still appeared very beneficial for most patients—but only for about six months. Then its benefits begin to wear off and side effects set in. Evaluation becomes complicated: the losses of benefits vary both over patients and over time; side effects also vary among patients and over time; prognosis is difficult both for benefits and for harmful side effects; overall evaluation must include prognoses for all those factors [Sacks 1973].

We must recognize that in many or most situations, we cannot obtain many and prolonged measurements; we must rely on models and other indicators of changes of effects. For example, in many cases if a single treatment *X* shows zero effects after a short but proper interval, one may assume the zero effect will continue into the future from *that* treatment. All the effects over time shown in Figure 3.6.4 are possible, but we hope that in most situations the decision lies chiefly between *A*1 and *B*1, or *A*2 and *B*2. Of course, other causes, as well as repeated treatments, may produce future effects. However, the combination of several treatments is too broad a subject for our brief discussion here.

The importance of taking into account seasonal factors in measurements is shown in

> the data of Marshall and Swan [1971] ... of average growth in height for two "cohorts" aged exactly 7.0 and 7.5 years at the beginning of the year. Over the one year period it is assumed that they both increase by 6.0 cm so that there has been no secular trend operating over the half year separating the cohorts. We also see that the fastest rate of growth is in the spring. Thus an estimate of average growth rate based on a period of less than a whole year will give a biased estimate. For example, using either cohort, if we estimate a growth rate based on the six-month period from January to July, we obtain a value of 7.2 cm per year, which is 20% too large. Likewise, if we base an estimate on the difference between the older cohort in April and the younger in July we obtain a value of 5.4 cm per year, which is 11% too small. [Goldstein 1979]

Not only seasonal but even diurnal variation may affect many bodily measurements. Several investigators have shown that measures of heights average about a half inch more in the morning, after sleeping flat, than in the evening after the compression of the spine due to the upright stance during the day. Substantive knowledge of any field is needed for planning the statistical design of studies over time. This brief section can do no more than call attention to this problem; and not only contradict the first quotation on instantaneous effects but also caution against complete reliance on single differences between pretests and posttests.

3.7 EVALUATION RESEARCH

Defining, describing, delimiting evaluation research (ER) to distinguish it from social research and research in general—yet covering its various special aspects for specific situations—poses a challenge. Also, because ER is a relatively new field, readers will come to it with very different backgrounds. Though earlier activities existed, 1955 may mark ER's formal birthday (in the International Social Science Bulletin 1955). In 1967 one could find some specific reports and a few manuals but no textbooks [Wright 1967]. By 1985 it had become a growth industry, with textbooks, courses, a journal, *Evaluation*, and an Evaluation Research Society.

> Two trends stand out in the modern attitude toward evaluation. First, evaluation has come to be expected as a regular accompaniment to rational social-action programs. Second, there has been a movement toward demanding more systematic, rigorous, and objective evidence of success. The application of social science techniques to the appraisal of social-action programs has come to be called evaluation research.
>
> A scientific approach to the assessment of a program's achievements is the hallmark of modern evaluation research. In this respect evaluation research resembles other kinds of social research in its concern for objectivity, reliability, and validity in the collection, analysis, and interpretation of data. But it can be distinguished as a special form of social research by its purpose and the conditions under which the research must be conducted. Both of these factors affect such components of the research process as study design and its translation into practice, allocation of research time and other resources, and the value or worth to be put upon the empirical findings. [Wright 1967]

Social research here includes epidemiology and health research [Susser 1975], but not laboratory or biological research on nonhumans.

ER then is a special kind of social research that concerns social programs and that involves decision making.

> Evaluations exist . . . to facilitate intelligent decision-making . . . if it does not improve the basis for decisions about the program and its competitors, then it loses its distinctive character as *evaluation* research and becomes simply social research. Most significant programs, we believe, are evaluated because some

decision maker wants help in figuring out what to do.... Perhaps the most common decision question of all is: Should we go on doing this or should we try something else, including doing nothing. [Edwards, Guttentag, and Snapper 1975]

This chapter is the best place to discuss ER because its problems of design (also of analysis) are essentially similar to the designs for comparisons in other kinds of social research. Yet ER deserves separate mention because its aims and scope are more specific, restricted, and clearly defined than those of social research in general. Here is my list of features that tend to distinguish ER from other kinds of social research, although no clear, thin line separates the two on any single feature.

1. ER involves specific social action programs. These programs may be local, national, or even international in scope; for example, some programs for contraception or against illiteracy have international organizations. Usually the programs need to be large enough to support the cost of ER designs that are good and large enough to have the accuracy and power to detect even modest changes. Most program improvements are modest, but they may have large social or financial consequences. On the other hand, programs that are obviously successful may not feel the need for ER.
2. ER typically involves cooperation between an agency (office) in charge of the social action program and a research team (institute) that undertakes the ER. Some separation of the team from the agency is needed: first to facilitate objectivity for the research, and second to enhance public perception and acceptance of the objectivity of the results of ER. Sometimes a grantor (source) of funds for ER may also be needed, separate from both the agency and the research team.

   Active cooperation between the two (or three) parties is typically vital. The agency has prime responsibility for the objectives of research, but the research team must take the lead in determining the methods and conduct of the research. In the selection of and cooperation from research sites, both freedom for the team and cooperation from the agency are often needed.
3. The choice of treatments and the observation of effects are both constrained by the practical needs of assessing the efficacy of the program under study.

   As a consequence [the team] has less freedom to select or reject certain independent, dependent, and intervening variables than [they] would have in studies designed to answer [their] own theoretically formulated questions, such as might be posed in basic social research. The concepts

> employed and their translation into measurable variables must be selected imaginatively but within the general framework set by the nature of the program being evaluated and its objectives. [Wright 1967]

These constrained objectives have several principal consequences. First, the specified combined program becomes the treatment (independent, predictor) variable, and its several components may be left untangled. For example, a new program of school instruction may be a complex of methods, teachers, and settings. Second, the measured effects must focus on the intended objectives as principal response (dependent, predictand) objectives. Nevertheless, it may be wise, and sometimes possible, to add long-range effects to mere short-range objectives. Furthermore, in addition to the planned objectives, some unintended consequences, both beneficial and especially harmful, should be anticipated and observed. Even further, the team should also be alert to detecting some entirely unanticipated consequences; "serendipity" may sometimes yield more interesting results than the expected main effects. Third, the program must be evaluated within its natural settings, under its operational conditions, and for its intended population. In all these respects it need not be subject to more severe tests of falsifiability which would go beyond the narrow confines of the program's specified conditions (7.6). Of course, passing such tests would add strength to the credibility, stability, and generality of the results of the ER.

4. Timing for ER often entails special problems. It is difficult to complete the research, followed by policy decisions based on its results, *before* starting a program, though this would be preferable for several reasons. Changing or dismantling an ongoing program would encounter obstacles even when it failed to yield results, because defects and criticism are bitter medicine for people involved in its operations. Furthermore, when an ongoing program covers the entire population, finding appropriate sites for "control" comparisons also poses problems.

   On the other hand, completing the ER and making consequent decisions for proposed new programs before their implementation entail delays and funding problems. Furthermore, administrative obstacles may hinder introducing the proposed program for an objectively randomized sample, which entails the problems of randomized treatments (1.4, 3.1).

   When the program is introduced on a partial basis, the research team should persuade the agency to build a good design for ER into the selection of sites and the timing for its introduction. Even if the program is being proposed for all sites, some delay for a sample of control (randomized) sites may be requested.

The need for hasty decisions for ER stands in contrast with the ideals of academic research for noncommitment, for universal laws, and for falsifiability (7.6). Faced with these problems and uncertainties, the statistical approach counsels reliance on probabilities and on repeated testing with ER to check the continued effectiveness of the program after change in the environment and in "side conditions."

5. "Evaluations exist to facilitate intelligent decision-making." Estimating the *effectiveness* of a program goes beyond measuring its effects. First, it is not enough to demonstrate the mere existence of effects against null hypotheses of zero effects with P values of tests of "significance." Nor is the explanatory power (R^2) of the treatment variables the main issue. Rather, the size of the effect becomes the central issue. Furthermore, often several effects must be measured, combined somehow—formally in an index or in a decision process or informally; often negative effects must be balanced against positive effects. In any case, the effect of the program must be weighed against a realistic and acceptable measure of its true total *cost*. The *cost effectiveness* of the program, a multipurpose function, must be assessed by several decision makers, and this may involve not only the program agency, but public bodies, experts, program recipients (present and future), and the public at large [Edwards, Guttentag, Snapper 1975].
6. The rules for publications of ER results may also differ from the accepted norms of academic social research. Reports to the agency typically must be hastened to fit early and fixed timetables. On the contrary, scientific reports to the public may be delayed, hindered, or prevented for reasons of confidentiality or lack of funds or lack of positive incentives.
7. The preceding needs of ER for constraints, decisions, speed, and restrictions often conflict with the usual norms of academic researchers. Yet independence and objectivity, both actual and reputed, are needed by teams doing ER (as noted in item 1). One desirable form of organization consists of separated research teams within public agencies, but with guaranteed independence and reputations. An early (1937) example was the Division of Program Surveys, established with independence, to evaluate the many programs of the U.S. Department of Agriculture [Converse 1986, Ch. 5]. A few other good examples can be found in the United States and elsewhere, but not often enough.

The growth of ER in the United States has also been reflected in the growth of institutes and centers that have been heavily involved in ER and in similar research, some of them attached to universities but some independent.

CHAPTER 4

Controls for Disturbing Variables

Not until attention has been effectually substituted for neglect as the general rule, will the statistics begin to show the merits of the particular methods of attention adopted. G B Shaw in Preface to The Doctors Dilemna.

. . . nothing improves the performance of an innovation more than the lack of controls. Muench's postulate in Mosteller's Experimentation and Innovations *(1977).*

4.1 CONTROL STRATEGIES

4.1A Four Controls Against Biases

Controls are needed in research designs for two essential reasons. One concerns the efficiency and economy of research projects and this may be stated in terms of reducing ("minimizing") either the variances or the costs of the projects. The second reason concerns the biases arising from disturbing variables in nonrandomized designs; it is this I shall emphasize in this chapter. In true (ideal) experiments those disturbing variables are all eliminated by the design, becoming either controlled *C* variables or randomized *R* variables. Reducing the effects from randomized *R* variables by increasing the effects of controlled *C* variables is the major concern for controls in experimental designs. However, additionally, in nonexperimental, nonrandomized observational studies some disturbing *D* variables may remain as factors that bias the observations and the comparisons of explanatory variables (Section 1.2). Here we shall emphasize the reduction of those biases, as we discuss measures for controlling the disturbing variables.

To deal with disturbing variables, one may resort to several alternative methods, depending on specific situations and resources. First, we may introduce appropriate controls into the sample design, i.e., into the selection

TABLE 4.1.1. Four Loci of Control for Disturbing Variables

1. Control in the selection of cases: allocation
2. Control in statistical analysis, e.g., weighting
3. Direct checks for biases in the study variables
4. Indirect checks for biases in disturbing variables

process; and the various methods for controls in the sample designs will concern us most in this chapter. Second, controls may also be introduced later in the statistical analysis to control those disturbing variables that we failed to control earlier in the selection design. In the absence of either kind of control, or if these controls are inadequate, one may resort to checks on possible biases; and these checks may take two forms. Thus, third, one may check study results against reliable base data which are directly comparable and sometimes, but not often, available; and then try to trace the connections between biases discovered in the results to some disturbing variables. For example, the results of biases in pre-election polls discovered after elections are usually "explained" *post hoc* with references to differences in timing, or in behavior, or in populations, or in other factors and their combinations. Such direct checks may be available for aggregate values of the study variables, and then used for rough, global evaluation of the quality of data; and perhaps also used for adjusting their effects on subclasses for which check data are unavailable. Fourth, instead of direct checks on the biases of study variables, one may try for indirect checks on differences in potentially disturbing variables in the sample against available population data (e.g., checks of standard demographic variables). (See Table 4.1.1.)

For comparing two or more treatment groups, each of the four kinds of controls would be used mostly for establishing the internal equivalence of the treatment groups; thus controls would be used to establish the "internal validity" of the treatment groups. The problems of "external validity," that is, of representation of the target population by the treatment groups, is a broader question for separate consideration (3.5).

4.1B Selection Control Versus Analysis Control

Although we noted four kinds of possible controls for controlling disturbing variables, we concentrate on controls in the selection process rather than later in the statistical analysis, because this is a book on design rather than analysis. This may be exemplified by control through allocation of sample cases contrasted with procedures for weighting sample results. In the literature of survey sampling this contrast is stated as stratification versus estimation. In experimental design the terms may be blocking versus analysis. Some of the following factors tend to favor one alternative and some the other. (See Table 4.1.2.)

Take first the nature (scaling) of the disturbing variables: categorical variables can be better controlled by selection, whereas analytical controls are easier for continuous, metric variables, and especially for linear relations and normal variables. For example, geographical variables like regions may be better controlled by selection, whereas income may be more easily controlled in the analysis, perhaps by covariance (4.6). This distinction still exists, but less clearly these days than formerly for two reasons. Analytical controls of categorical data have been made less forbidding by statistical advances in both computing and statistics (4.6). Also, there is less confidence nowadays in linear models and in the normality of continuous metric data.

Second, arbitrary selection and allocation of sample size appears more feasible in experimental designs than in large-scale survey sampling, where it is more costly and difficult to manipulate sampling rates and numbers. Survey samples, therefore, may need and use analytical controls more often.

Third, model-dependent research relies more readily on allocations and on selections based on models than does population-bound research, which aims to represent target populations (1.8). Perhaps such reliance on models versus populations is mostly a justification which follows that process of choosing a feasible research design rather than precedes it, as it should. But if the influence of feasibilities can lead to overreliance on models, we have another case of wishes fathering thoughts.

Fourth, control by selection may seem rather simple when planning for only a few and only relatively simple statistics. But for more complicated and more numerous statistical results, additional controls for disturbing variables would be more difficult. Nevertheless, control by selection may prove to be wise foresight, especially for multipurpose surveys (7.3).

Fifth, control by selection requires that information on all sampling units be available. But control by reweighting may be used with information available for only the sampled units and the population aggregates (4.7).

Sixth, when controls through analysis result in grossly unequal weights, the variances of estimates can greatly increase (7.4). Then controls by selection and design should be preferred.

Seventh, on the other hand, when control by design would involve expensive screening procedures in the selection process, then controls by weighting may be more economical. This choice is illustrated by contrasting matched cases (4.3) with matched subclasses (4.4) and with standardization (4.5). For this and for the fifth reason, religious identification, not available

TABLE 4.1.2. Strategies for Controls by Selection Versus Analysis

For Controls by Selection	For Controls by Weighting in Analysis
1. Categorical data	Continuous, linear, normal variables
2. Experimental designs	Survey sampling
3. Model-dependent research	Population-bound research
4. Few, simple statistics	Complex, multipurpose analysis
5. Data on whole population	Reweighting of sampled units
6. Highly unequal domains	Small inequalities of weights
7. Inexpensive screening	Expensive screening

for the population and difficult to screen, may be better controlled by analysis.

The conflicting directions of all these criteria make it abundantly clear that we should not be swayed by only a single consideration and that we need informed judgment to guide our choice between the two alternative methods of control over disturbing variables.

I also owe a two-edged explanation for this chapter and especially for Section 5 on weighting and standardization. First, these cover only a fraction of the tools available for statistical controls by analyses; but we cannot cover those vast fields here. Second, the reasons for including weighting and standardization here (although they belong to analysis rather than design) are that they are the simplest and the most common tools, yet they are neglected in most of the modern statistics textbooks of today (though they had appeared in some older texts).

4.1C Choosing Variables for Selection Control

For most research objectives, such a profusion of potentially disturbing variables exists that we need some guidance for choosing a few of them for controls in the selection process. Although we may read about "controlling for all disturbing variables," that advice is neither reasonable nor practical. There are just too many potentially disturbing extraneous variables that may affect the explanatory variables in most research situations, especially in nonexperimental designs. Practical and economic considerations force the researcher to confine controls to only a small number of variables (4.1D). We should choose to control those that seem most important on theoretical grounds, or such as appear most useful on empirical evidence from past research. These two criteria should ideally direct us to the same disturbing variables, but not necessarily in the real world of imperfect theories; and we should remember both criteria. For example, yearly income may seem theoretically like a good control for socioeconomic status, but education has often been found to be a better predictor, because it can be obtained with greater accuracy and with more stability over time.

Before discussing their desirability and relevance, we should also consider briefly the availability of data for controlling disturbing variables. Good information about availability is a practical asset for any research project, though the search should be guided by a framework of theoretical concepts. Availability may be more complex than a mere yes/no dichotomy: some variables may be easily accessible on public tapes, others may be obtainable only at great cost, and many at too great a cost or not at all. For control in the selection, the disturbing variables must be available for all units at each proper stage of selection and over the whole population frame. For example, information for all high schools of a state would be needed to select two sets of schools for treatment and control, if we want the inference from the comparison to have statistical (external) validity to all the state's schools. However, a small proportion of "unknowns" may be tolerated either as a "miscellaneous" class or as a deliberate "exclusion" from the frame population (2.1).

On the other hand, control in the statistical analysis can use information available on the sampled units only, when these are combined with aggregate population data in ratio estimates (4.7). Reweighting the data may be used to correct for differences in disturbing variables between treatment groups, because the data were not generally available and had to be obtained during the research itself. The researcher can also deal in the analysis with other disturbing variables, which were not controlled in the selection process simply because they were overlooked, ignored, missed, or unknown during the selection process.

In survey sampling one also faces similar restrictions when auxiliary (ancillary) data are not available for all units in the population. However, instead of confining the data collection to the sample, gains can sometimes be made by obtaining the auxiliary data from a larger sample, even though this can be much smaller than the population. This "screening" operation is known as the first phase of "two-phase" (or multiphase, or double) sampling [Kish 1965, 12.1; Cochran 1977, 12]. This method can also be applied to observational studies and experiments.

Let us now consider how to choose, for control by selection, some of the variables from a larger set of potentially disturbing *D* variables. It may be feasible and worthwhile to begin with a theoretical outline of all potential *D* variables, whether available or not. A list of the unavailable potential D variables may impart salutary caution to researchers and also goad them perhaps to search deeper. The list of the available D variables may be divided into several classes.

1. For some D variables control seems clearly (*prima facie*) necessary and possible.

2. For other variables, controls may seem unnecessary or unimportant (perhaps tentatively and probably relatively) in comparison with other candidates.

3. Still others may be subjected to formal tests to determine whether they should be among the controlled (1) or the uncontrolled (2). These "formal tests" would be more demanding presumably than the informal process used to separate classes 1 and 2 above. We may test or probe for one of two kinds of differences between treatments in the disturbing variables: either in their frequency distributions or in their values along that distribution. Happily either one of these two probes will be sufficient, because uniformity over one type is sufficient for lack of disturbing bias, and because both kinds of probes may be too difficult. First, if the frequency distributions of the disturbing variable are similar for both (or all) treatments, then even possible differences in effects do not bias the average effect. Second, if the effects of both (all) treatments are uniform over the frequency distributions, then differences in frequency distributions between treatments do not bias the average effect (4.5B). This is clearly shown for the special case when the disturbing variable D has a linear regression on the study variable Y, by Cochran [1965, p. 3; 1983, 5.1].

For example, suppose we wonder about the effects of age as a disturbing variable when testing the difference between two aspirins. If the age distributions are similar for the two treatment groups, the possible age differences in effects would not bias the average difference of the two aspirins. If the effects are similar for all ages, age differences between the groups would also not bias the average difference. If we cannot be satisfied with either kind of uniformity, a potential bias may lurk in the disturbing variable, age the example.

Even if the differences of D classes are marked, and even if "statistically significant," they may have only negligible effects on the overall means and on their differences if the extreme classes are relatively small portions of the entire populations. Furthermore, even in the presence of both kinds of differences, the net effective bias may still vanish from the difference (comparison) of the means if the diverse effects within classes happen to cancel each other (4.5B). Such canceling effects are not rare events, but they are rarely predictable or reliable.

Some variables that can be controlled may be deliberately included (combined) within the predictor variable for empirical reasons (for greater realism), although they could be considered extraneous on some theoretical grounds. Suppose, for example, that we compare two techniques of classroom teaching: Should the motivations, attitudes, and qualifications of teachers be considered jointly with the techniques, as their necessary implementations? On the contrary, should they be considered as disturbing variables, whose effects we would separate, control, and measure? What about classroom sizes and class organization: are they *D* variables or part of the predictor? Similar questions about predictor and disturbing variables arise in most actual situations. Variables in research seldom appear as obvious, pure, unidimensional variables—like mass and time in physics. Many measured variables actually represent combinations or vectors of several basic variables that potentially could have been separated—either operationally or only conceptually. But they are often treated jointly as a single variable, because of the restrictions or needs of the research situation. Thus the definitions of predictor variables may include those that in other contexts would be considered *D* variables.

4.1D Numbers and Classes of Control Variables

We may still be left with a longer list of available *D* variables, all of which we would like to control with selection, than we can afford to use. Faced with a number d of disturbing variables, if we want c cells for each, we would end up with c^d cells, which can be enormously large even for small values of c and d. For unequal numbers c_i, a similar picture emerges for the product $\Pi_i c_i$ of d terms. Though we may combine some cells that are small or empty, the total number of cells can still be much too large for controlling. Such situations often produce conflicts about which variables to control and how many cells to use for each. Fortunately, the statistical methods of sampling and regressions offer some answers worth borrowing.

1. It is better to use several variables, each with only three to five classes, than many classes from one or two variables. As few as three to five classes suffice to yield most (80 to 90 percent) of the control available from any single variable (Table 4.6.1). Thus for a limited number of cells, one can obtain better overall control with few cells each for more *D* variables than with many cells each for fewer *D* variables. This strategy resembles the choices for predictor variables in regressions and for stratification in surveys [Cochran 1968; Kish and Anderson 1978]. It may be modified to use more (say, six) classes for one important variable and fewer (say, three) for the less important (7.3). These general rules obviously need care in specific applications.

2. To choose a small number of control variables from a larger number of

D variables, one should spread those few choices among variables that are not highly interrelated. High correlations between control variables would reduce the contribution of each to the aggregate control. This problem and strategy resemble the choices both of predictor variables for regressions and of stratifying variables in sampling [Anderson, Kish, and Cornell 1980].

3. Recommendations 1 and 2 hold with even greater force and efficacy when we design multipurpose studies for several study variables. Then it becomes even more likely that we need to control for several disturbing variables rather than for only one or two "optimal" variables and that the controls need to be "spread" rather than interrelated (7.3) [Kish and Anderson 1978].

4. Sampling methods can also help to locate the class boundaries for variables to obtain increased control: "optimal stratification" calls for boundaries that yield classes with (approximately) equal values of $W_h\sigma_h$, the product of the size and variability within classes [Anderson, Kish, and Cornell 1976]. That strategy tends to compromise between equal sizes and equal deviations for the classes, which tend to be inversely related in skewed frequency distributions. This can be approximated closely with continuous variables and roughly for discrete classes by combining some of them. But such opportunities for creating class boundaries are lacking for those D variables that come fixed in a few classes, and when the boundaries are determined by substantive/theoretical considerations.

5. Perhaps several (or even many) D variables may be combined with factor analysis or clustering methods into a few variables. This can probably be done better for sets of related variables. These methods have been advocated for years for stratified sampling, but they have not yet been shown to yield better results than those given above.

4.2 ANALYSIS IN SEPARATE SUBCLASSES

For this most obvious and common method of control the sample is separated into subclasses according to some disturbing variable, and means and their differences $(\bar{y}_{ai} - \bar{y}_{bi})$ are then computed for the separate subclasses, denoted by i. The subclass differences can then be compared with the overall difference $(\bar{y}_a - \bar{y}_b)$; studies of these relationships through *inspection* and *introspection* is perhaps the most common method of survey analysis.

Subclass analysis is also known as *domain analysis*, because sample subclasses represent corresponding domains of the population (2.3). This method is used not only for control of disturbing D variables, but more prominently for finding and interpreting relations among explanatory E

variables (predictors and predictands). Explanatory analysis is an important, vast, and separate subject we cannot hope to cover here. It is covered briefly and well by Moser and Kalton [1971, 17.4], with references to fuller treatments [Hyman 1955; Rosenberg 1968; O'Muircheartaigh and Payne 1977].

Our discussion here must concentrate on the use of subclasses i for discovering possible effects of disturbing factors on the overall predictand response $(\bar{y}_a - \bar{y}_b)$ attributed to the predictor treatments (a, b). We concentrate on the use of subclass differences $(\bar{y}_{ai} - \bar{y}_{bi})$ to test the overall difference $(\bar{y}_a - \bar{y}_b)$ against possible effects of potentially disturbing D variables, denoted by i. We only mention incidentally some deep problems that can arise in such analysis. First, when subclass differences are found that seem large and interesting, how far should their causes be explored and explained? To exploit the discovered differences, one can transform these disturbing factors into explanatory factors. Such transformation and expansion of research objectives are especially tempting when large subclass differences $(\bar{y}_{ai} - \bar{y}_{bi})$ are found despite small overall differences $(\bar{y}_a - \bar{y}_b)$. Second, subclass comparisons lead to questions of representation. Third, they lead to questions of randomization and of the balance of treatments over subjects (elements).

Subclass means $\bar{y}_i$ represent the simplest forms of control, and the comparison $(\bar{y}_i - \bar{y}_j)$ can denote subclasses from the same control variable. For example, it could represent age-specific birthrates for two age groups (i, j) of mothers. However, $(\bar{y}_a - \bar{y}_b)$ can also denote a more complex concept: the difference between two treatments, hence of two populations; for example, the difference in birthrates in two populations exposed to two contrasting treatments of contraception (a and b). Subclass control is denoted by $(\bar{y}_{ai} - \bar{y}_{bi})$, for example, age-specific (i) differences as controls for the overall difference between birthrates comparing the two methods (a, b) of contraception.

Proportions p denote a special form of means, which are very common for the counts of categorical data in social (and other) research. The values of $p = y/n$ can range only from 0 to 1; and in $p_{ai} = \bar{y}_{ai}/n_{ai}$, the $\bar{y}_{ai}$ range only from 0 to n_{ai}, because they are based on elemental categorical counts of 0 and 1. However, births per woman in integral counts of births (0, 1, 2, 3, . . .) have wider ranges, and other data (weight, income, net savings) wider still. The proportions $(p_a - p_b)$ can also denote the difference in death rates from two treatments (a, b), for example, and $(p_{ai} - p_{bi})$ the age-specific controls (i) for that difference in death rates. Furthermore, instead of differences, the effects of the treatments can also be measured first by the ratios p_a/p_b and then by the ratios p_{ai}/p_{bi} in the control subclasses i. When the proportions can be viewed as probabilities, the ratios are called *likelihood ratios*. In categorical data analysis, furthermore, *odds ratios* are often used, which we can symbolize with $[p_{ai}/(1 - p_{ai})]/[p_{bi}/(1 - p_{bi})]$.

The sources of the data for the difference $(\bar{y}_a - \bar{y}_b)$ can vary greatly. The treatments a and b may be assigned by randomization in ideal experiments. Or they may be separated and controlled in observational studies. Or the a and b may be subclasses from data of a sample survey. For example, the $(\bar{y}_a - \bar{y}_b)$ may denote female–male differences in a survey study of the "gender gap" in voting behavior, or in occupations, salaries, death rates, etc. Then the $(\bar{y}_{ai} - \bar{y}_{bi})$ denote female–male differences computed within subgroups to control for disturbing variables (e.g. age or education).

We cannot explore here questions about how to choose disturbing variables for controls by subclass analysis. Clearly, more than one disturbing variable will be tested in most situations. Often two or more will be used jointly; in such cases the subscripts i in $(p_{ai} - p_{bi})$ denote subclasses created by several variables. Multivariate controls should yield more powerful tests for reasons similar to those for designs (4.1C), although different kinds of variables should be preferred for control by analysis (4.1B). Yet control with subclasses has a numerical limitation in common with controls in selection: The number of variables and classes cannot be large, because of bounds imposed by the sample sizes, and also by limits on the complexity of analyses. To use more disturbing variables and also more classes, all simultaneously, one may resort to more complex multivariate techniques, including categorical data analysis (4.6).

Let me juxtapose two entirely different aims for subclass analysis: one to find predictor variables to "explain" the relations $(\bar{y}_a - \bar{y}_b)$, the other to test the relations $(\bar{y}_a - \bar{y}_b)$ against potentially disturbing variables. When subclasses are used for causal analysis, i.e., for discovery of explanatory factors for a difference $(p_a - p_b)$ of a dichotomous variable, the research aims to find $(p_{ai} - p_{bi})$ with maximal discriminatory power. The ultimate, which is seldom even approached, would be a deterministic model: when the subclass values p_{ai} and p_{bi} would be either 0 or 1, and the $(p_{ai} - p_{bi})$ either $+1$ or -1. That is, the subclasses i would explain entirely when and where the treatments a or b, respectively, are either entirely successful (0) or unsuccessful (1). Also for more general variables we should aim to increase (maximize) the differences among the $(\bar{y}_{ai} - \bar{y}_{bi})$ for different i.

On the other hand, in using subclasses for controlling disturbing variables the principal aim consists in testing the overall difference $(\bar{y}_a - \bar{y}_b)$ against subclass differences $(\bar{y}_{ai} - \bar{y}_{bi})$. To the degree that the $(\bar{y}_{ai} - \bar{y}_{bi})$ appear similar to the $(\bar{y}_a - \bar{y}_b)$, this overall difference survived the tests of falsification by the potentially disturbing factors represented by the subclasses i. Surviving the severest tests of falsification yields the strongest confirmation of the treatment effects in $(\bar{y}_a - \bar{y}_b)$. With this use of Popper's falsification method, confirmation is approached to the degree that the $(\bar{y}_{ai} - \bar{y}_{bi})$ resemble the $(\bar{y}_a - \bar{y}_b)$ under the severest testing with the subclasses i (7.6). In other words, as we inspect different subclasses, when controlling for disturbing variables, we hope to find similar $(\bar{y}_{ai} - \bar{y}_{bi})$, whereas for explanatory E variables we aim to increase those differences.

We are concerned mainly here with subclasses for controls, hence this sharp contrast. This conceptual contrast is probably novel; and it is not formulated sharply in the minds of researchers. However, this distinction of the two aims of subclass analysis may also be applied to other research objectives, to more than two treatments, and to single means $\bar{y}$ or p, etc. In all these cases we should distinguish the search for predictors with "explanatory powers" from the tests of those explanations against potentially disturbing variables.

"*Simpson's paradox*" is a dramatic device for portraying problems with subclasses. We clearly have $p_a = \Sigma w_{ai} p_{ai}$ for nonnegative relative weights $(0 \leq w_{ai} \leq 1)$ that add to $\Sigma w_{ai} = w_a = 1$. Often the weights denote numbers of cases and $w_{ai} = n_{ai}/n_a$; then $p_a = \Sigma(n_{ai}/n_a)(y_{ai}/n_{ai}) = \Sigma y_{ai}/n_a = y_a/n_a$. The p_a is an average of the p_{ai}; hence it must lie between their extremes (the largest and the smallest of the p_{ai}), not outside them. Similarly, p_b, as an average of the p_{bi}, must lie between their extremes. Simpson's paradox states that it is possible to have $(p_a - p_b) \geq 0$ in spite of having $(p_{ai} - p_{bi}) \leq 0$ for all the i. For example, one may have a new treatment (a) better in each of two clinics (1, 2), so that $(p_{a1} - p_{b1}) < 0$ and $(p_{a2} - p_{b2}) < 0$, yet the old treatment (b) better overall $(p_a - p_b) > 0$ where p stands for failure or death. This can happen if one clinic that uses more of the new treatment also gets more of the high-risk patients [Blyth 1972]. In that case the subclasses (clinics) reveal the better new treatment (assuming random assignments within clinics), which remains hidden in the overall difference.

On the other hand, the overall difference presents more meaningful results in an example of income subclasses:

> Between 1974 and 1978, the tax rate *decreased* in each income category, yet the overall tax rate *increased* from 14.1 percent to 15.2 percent. Again, the overall rates are weighted averages, with the tax rate for each category weighted by that category's proportion of total income. Because of the inflation, in 1978 there were relatively more persons and consequently relatively more taxable dollars assigned to the higher income (i.e., higher tax rate) brackets. [Wagner 1982]

But if we were ignorant of the inflation as the real explanation, we might have accepted the decreases within categories as more meaningful than the overall increase. For other good examples see Cohen [1986] and Bickel, Hammel, and O'Connell [1975]. Cohen writes:

> Though every age-specific death rate in Sweden is lower than the corresponding age-specific death rate in Costa Rica, the crude death rate of Sweden exceeds that of Costa Rica. [Because] The Costa Rican population has a much higher proportion of young individuals, whose death rates are less than those of old individuals in either Costa Rica or Sweden.

"Simpson's paradox" concerns reversals of the signs of differences between the $(p_a - p_b)$ and the $(p_{ai} - p_{bi})$, and "collapsing tables" may be a better name for the general phenomena [Fienberg 1977, 3.8; Yule and Kendall 1965, Ch. 2]. The "paradox" presents only a special case of more general divergences between the overall results $(\bar{y}_a - \bar{y}_b)$ and the $(\bar{y}_{ai} - \bar{y}_{bi})$; or even between the overall mean $\bar{y}$ and subclass means $\bar{y}_i$; or other divergences between overall and subclass results. We refer here to divergences beyond mere sampling fluctuations (to be investigated with sampling errors), hence divergences similar to those one could presumably find in the population also.

Biases investigated with subclasses of disturbing variables tend to vanish when the differences in either the weights $(w_{ai} - w_{bi})$ or the subclass means $(\bar{y}_{ai} - \bar{y}_{bi})$ tend to zero (4.5B). Investigations of reasons for differences $(\bar{y}_{ai} - \bar{y}_{bi})$ should lead to substantive analysis of relations between the disturbing variables and the explanatory variables. On the other hand, differences in $(w_{ai} - w_{bi})$ should lead to questions about control. But control through separate subclass analysis evades the task of yielding a single combined estimate of the relationship $(\bar{y}_a - \bar{y}_b)$. This task is accomplished in later sections that present alternative methods for eliminating biases from specified subclasses by equalizing their weights, $w_{ai} = w_{bi}$. This can be done either in the selection design (4.3 and 4.4) or by reweighting in the analysis by standardization (4.5).

Restriction to a single subclass may be regarded as a special, or extreme, or degenerate case of control by subclasses: To avoid effects from disturbing variable(s) the researcher deliberately restricts the scope of the research to a single subclass of the disturbing variable(s). Such restrictions may be justified if imposed by stringent economic demands and if they bring commensurate savings. These considerations lead, for example, to community studies restricted to a single local area or to a single social organization, school, or firm (3.1A). But such units seldom actually contain a single, pure type of population that is free from disturbing factors. Actual pure types are exemplified instead by the constants that specify highly standardized materials and conditions for some types of physical and chemical measurements; but then in practical applications, allowance must be made for imperfect materials and standardizations. Furthermore, such control is neither practicable nor desirable in social research (also seldom in medical or biological research) because its subjects (humans, animals) cannot be standardized. The penalty for restricting research is often a drastic narrowing of the inferential scope of its results; this was opposed and countered by the *factorial designs* of R A Fisher (1.3).

Limitations imposed by "pure" types of subclasses are more severe than those from local studies (3.1), because these latter may cover a broader population that is closer to the population(s) of inference. Hence for inferences from pure types the need for replications with other pure types becomes even greater. By the same token the opportunities for testing against possible falsification can also be increased by choosing very different "pure" types for the several subclasses (7.6).

The situation is quite different when a subclass is selected for convenience in sampling or in data collection. Good examples are two studies dealing with cohorts; four birthdates were used for longitudinal studies of $4/365 \simeq 1$ percent sample censuses in the United Kingdom [Douglas and Blomfield 1956]; and a single week's birth cohort is followed in Sweden [Janson 1984]. For studies of schoolchildren, specified grades in schools have been used conveniently; grades are more convenient than exact years of age would be, and they may be as appropriate for defining the population of the subclass. Convenience and appropriate definition may coincide; for example, one study was confined to women giving births to their third babies in the hospitals of Detroit [Freedman, Thornton, and Camburn 1980].

4.3 SELECTING MATCHED UNITS

This method will first be described in its extreme form of case-by-case matching, which is perhaps the most common method; later some modifications will be added. Two categories, **a** and **b**, of the predictor variable are used to establish the two subpopulations to be compared. Similar subclasses i are formed in both subpopulations in accord with an appropriate subdivision of the control variables. One unit is selected from each subclass for both categories, thus $n_{ai} = n_{bi} = 1$. The differences $d_i = (\bar{y}_{ai} - \bar{y}_{bi})$ in the subclasses become the basis for data analysis, whose problems we cannot explore. We merely point to problems of inference, especially the need for superpopulation models when an entire small subpopulation is used for the **a** treatment.

For a specific example take the study "Unemployment and Migration in the Depression (1930–1935)" [Freedman and Hawley 1949]. The data came from schedules of the month-by-month history of the Michigan Population and Unemployment Census of 1935. The sample of group **a**

> consists of all those white male migrants to Flint or Grand Rapids from other places in Michigan, who were at least 25 years old at the time of migration to the cities. . . . Flint and Grand Rapids were selected as the destination points to be studied, because they are very different both with respect to population history and economic base. Therefore, it was hoped that studies of the migrants to each place might be treated as separate "experiments." Agreement to the findings for the two cities should give them greater validity.

This is a clear statement for "internal validity" in local studies (3.1B) and for falsifiability (7.6). For groups **b** *and* **b**′

> each migrant was "matched" with a "control" non-migrant at the place from which he came and another at the place to which he moved (either Grand Rapids or Flint). The characteristics used for matching were age (within 3 years), occupation (in terms of the major census socioeconomic groups), occupational history (in terms of change between socioeconomic classes) education (within 2 years of school achievement), and marital status.

The two control groups (b and b') for the one treatment group (a) form the basis of two studies to find "that the differential in unemployment occurs after migration, not before . . . that in a depression migrants tend to be at a disadvantage in the new labor market to which they move." Furthermore, these two comparisons were part of a larger "series of studies of the relationship between migrant status and a number of other characteristics (education, occupation, occupational mobility)."

Several aspects of this example are common to case-by-case matching. The data are taken from existing schedules; detailed matching by field collection would entail prohibitively high costs. Often the "experimental" treatment group **a** is small compared with the numbers available for "control" treatment **b** and compared with the number of groups used for selection controls **i**. (We just used the word *control* in two different senses.) For those two reasons we can assume that a single case for each "experimental" subclass is common, thus $n_{ai} = 1$. Another reason is the large number of subclasses formed when the controlled variables are numerous, five in our example. But the n_{bi} for the "control" treatment are often more numerous; then questions about methods for selecting among them arise, especially if it is too difficult to sort out separately all the n_{bi} candidates for each subclass. The example illustrates a common method:

> The procedure was essentially to enter the schedules at a point determined by a system of random numbers, examine the schedules serially until a match was found, then to select a new starting point and to repeat the process. Several alternative methods involving selection of all possible eligible matches and random selection among them were dismissed as prohibitively time-consuming.

The authors recognized the subtle source of potential bias in selecting the "next eligible" listing [Kish 1965, Sec. 11.1], but found no actual bias in the results.

Even when the "control" cases are more numerous, cells with unmatched cases with $n_{bi} = 0$ often occur, owing both to sampling and to structural variations in the **b** subpopulation. "Of the 360 migrants to Flint 312, or 87 percent were matched at the destination and 296, or 82 percent at the source-points. Of the 186 migrants to Grand Rapids 170, or 92 percent were matched at the destination and 149, or 80 percent at the source points." We note here that the proportion matched can be increased, even to 100 percent, by relaxing controls to create larger cells whenever $n_{bi} = 0$. On the contrary, it may be possible to obtain "better" matches whenever the cell sizes are greater than 1 by increasing the stringency of controls.

The case-by-case matching of units represents an extreme emphasis on "better" control of individual cases with multivariate matching. On the other hand, with looser fitting, with fewer subclass cells, and with larger samples we may achieve stricter random selections and lower variances (4.4).

Methods for case-by-case matching of units have been developed recently in other areas, together with procedures for using computers for data from tapes or disks. Methods have been developed in epidemiological and medical research and general treatments are given in chapters on "matching" in several books [Cochran 1983, Ch. 5; Anderson et al. 1980, Ch. 6; Kleinbaum, Kupper, and Morgenstern 1982, Ch. 18; Fleiss 1973, Ch. 8].

Imputation for missing items has been developed mostly at the U.S. Census Bureau and at Statistics Canada. The procedures impute missing items for class 1 cases where those items were omitted from class 2 cases that have them. The procedures use multivariate subclass matching of schedules, to impute to a minority subpopulation with missing values (1) the values from a larger subpopulation with the needed data (2). The multivariate, case-by-case method has both aims and procedures similar to our present needs, and researchers can benefit, I believe, from the large, new literature on machine imputation [Kalton 1983; Rubin 1978].

High-speed computers can be used to find easily the "next eligible" matching case from the b subpopulation with a "hot deck" procedure for imputation. Instead of using the "next" case on the ordered list, the procedure could be modified to select a "better fit" with additional variables.

Other procedures ("modified hot deck" imputation) can identify with fast computers the entire set n_{bi} of matches for each missing item and select one randomly, thus eliminating a possible source of potential bias. Furthermore, with all n_{bi} potential matches identified, it may be possible to use several (or all?) of them for matching each missing **a** case, thereby decreasing drastically the variance of the comparisons (of the d_i values) [Kalton and Kish 1981; Rubin 1978]. However, this extension of the method fits better into Section 4.4.

Another extension needs only a brief mention: the method can be used not only for two subpopulations a and b, but also for more: c, d, e, f, etc. In our example the migrants (a) were matched with nonmigrants at both the departure-source (b) and at the arrival-destination (b'); and other reports from that study also made additional comparisons.

Matching of data files presents some similarities, but fewer procedures directly applicable to our present purposes. There are two basic types of matching for files: "exact" and "statistical" [Radner 1979; 1980]. For exact matching, some linkages of data for the same units (e.g., persons, firms) are sought; these require using identifiers such as names, addresses, Social Security numbers. For statistical matching, linkages for "similar" units are sought and expected; and files based on samples, with only few accidental units (or none) in common, are the rule. Statistical linkage uses "similar" characteristics, based on several relevant (related to the study) variables, rather than unique identifiers, and thus resembles case-by-case matching in some ways [Radner 1979, 1980; Rodgers 1982].

4.4 MATCHED SUBCLASSES

Section 4.3 dealt with case-by-case matching, where one case with "control" treatment (b) is selected from the $n_{bi} \geq 1$ available to match each single (a) treatment, so that $n_{ai} = n_{bi} = 1$ for the analysis. At the other extreme, Section 4.5 describes methods of standardization for utilizing all available $n_{ai} \geq 1$ and $n_{bi} \geq 1$ cases, adjusting the numbers by weighting so that $w_{ai} = w_{bi}$. Here we explore briefly some possible procedures that lie between those two extremes, and these compromises also illuminate the extremes.

Case-by-case methods are better suited to small samples, especially where $n_a = \Sigma n_{ai}$ is small. There are many subclass cells created, because emphasis is on close fits of cases to reduce bias, and for most cells the n_{ai} are either 1 or 0. In contrast, standardization can use large samples, with relatively few controls and few subclass cells; therefore sorting out all n_{ai} and n_{bi} for all subclass cells (i) is feasible; sampling may be introduced if samples are larger than needed.

For matched subclasses, along with bias reduction, sampling variation should also be considered for choosing among the several alternative designs briefly noted below.

1. *Use $n'_{bi} > 1$ to reduce sampling variation.* This may be applied even when all or most $n_{ai} = 1$. It is possible either to produce an average $\bar{y}_{bi}$ for all n'_{bi} values of the y_{ai} values used, or to compute separate values of $d_i = (y_{ai} - y_{bi})$ for each [Kalton and Kish 1981]. Both methods should yield the same mean difference, but other analyses could differ. A conflict arises: to use all available n_{bi} cases of the "control" treatment b in the ith cell would minimize the variance, but using a constant k_b cases for each cell (where $n_{bi} \geq k_b \geq 1$ and $n_{ai} = 1$) may simplify the statistical analysis and computing variances. The presence of several b cases within each cell will also contribute to knowledge of within cell variations. Maintaining a constant $k_b \geq 1$ may involve replicating a few cases in cells where the actual $n_{bi} < k_b$.
2. *Use constant sizes for both a and b*, with $k_b \geq k_a \geq 1$. For simplicity's sake this may be adopted when the supply of the n_{ai} and n_{bi} is so large that we can afford to sample from them. Equal constant subclass sizes $k_a = k_b$ may seem the simplest. However, $k_b > k_a$ may seem advisable (1) when the supply of b cases is larger, or (2) when the handling of b cases is less expensive. This latter would accord with principles of "optimal allocation" (7.3E).
3. *Use equal sizes within subclass cells: $n_{ai} = n'_{bi} = n_i$.* When the n_a for treatment **a** is small, the n_{ai} will determine the sizes of the subclass cells. The $n'_{bi} = n_{ai}$ cases are selected at random from the $n_{bi} > n'_{bi}$ cases available in the ith cell to match the n_{ai} cases. In this respect it resembles case-by-case procedures, except that now the n_{ai} cases in the cells are computed together. It also implies weighting the overall means by sizes n_{ai} available for the "experimental" treatment **a** as weights, and it may be viewed as a special case of standardization procedures in Section 4.5. This can be modified to $n'_{bi} = kn_{ai}$ with $k \geq 1$. In other cases n_{bi}, or some other cell weights, may be adapted as the standard.

Matching by selection wastes the reservoir of cases and the effort required for selection. These two kinds of waste become large when a relatively rare treatment **a** is matched against a treatment b from a large total pool of cases. If that pool is very large (e.g., a population census), the wasting of cases may not be too important, but the effort and expense needed for selection may still be exorbitant. If both treatments a and b are large, the search for matches would be even more difficult and consuming. That is why I placed case-by-case matching into the context of searching records. With modern computers for searching tapes, such tasks are becoming increasingly possible.

On the other hand, field screening to find multivariate matches would be much more difficult and more expensive. Therefore, it is seldom done, though theoretically possible. That possibility would be increased with marginal matching of univariate variables [Yates 1949, 3.4].

Reducing the dimensions of matching to single or fewer variables may also be needed for sampling from records and tapes. This need arises when each of several records contains only some of the variables and the matching must be along several margins of the full multivariate structure. Diverse literature exists in several fields that are only thinly related [Deming 1943, Ch. 7; Purcell and Kish 1980; Rodgers 1984; Radner et al. 1980].

4.5 STANDARDIZATION: ADJUSTMENT BY WEIGHTING INDEXES

4.5A Weighting Versus Matching by Elimination

In Sections 4.3 and 4.4 the sizes of disturbing subclasses were controlled by equalizing them in the selection process, so that $n'_{ai} = n'_{bi}$ cases were selected for analysis. Here we propose to retain all the available n_{ai} and n_{bi} cases, but equalize these subclass sizes by using common weights w_i for both treatments **a** and **b** within subclasses i, with $\Sigma\, w_i = 1$. Thus by computing $\Sigma\, w_i(\bar{y}_{ai} - \bar{y}_{bi})$ instead of $\Sigma\, w_{ai}\bar{y}_{ai} - \Sigma\, w_{bi}\bar{y}_{bi}$, the effects of different subclass sizes for **a** and **b** are removed from the reweighted estimates of the differences between treatments.

Equalizing the sizes of subclasses by weighting serves as a direct substitute for equalizing by subselection. That is one reason for including standardization here, whereas other methods of statistical analysis have been excluded generally from this book. Furthermore, standardization is not covered in most textbooks on statistics, although it can be found in some textbooks on economic statistics and in the demographic literature [Shryock and Siegel 1973; Kitagawa 1955, 1964; Yule and Kendall 1965, Ch. 25; Hill 1961, Ch. 1]. Also, this section can well serve to reveal the basic nature of adjustments for disturbing variables.

Matching by selection wastes the reservoir of cases; it also increases the efforts required for selection. Equalization by weighting reduces those losses, but the weighting process requires more computation and more complex analysis. The "losses" here refer to increases in the variances that result when the number of cases matched is n'_{bi} from a larger poor of n_{bi} available cases; variance is increased roughly in the ratio of n_{bi}/n'_{bi}, or n_{bi}/n_{ai} in case-by-case matching. In this comparison we consider only the variances, because we assume that weighting versus selection for controls have similar effects for reducing biases from disturbing variables.

Those losses are reduced when all the available $n_b = \Sigma n_{bi}$ and all the n_a cases are used in n_a pairs. But the losses are not entirely eliminated, because the weighted estimates tend to have higher variances than a similar number $(n_a + n_b)$ of cases from a matched selection of $(n_a + n_b)/2$ pairs would have. Weighted samples tend to have somewhat higher variances per sample case used than unweighted samples (7.5). But this increase in the variance due to weighting is not as serious as that arising from elimination.

Furthermore weighting also requires more computations and more complex handling. These are facilitated by modern computers, but "human errors" still lurk in the practical aspects of handling weights. Theoretical difficulties can also be expected with more complex estimates, including sampling errors, tests of significance, and analytical statistics (7.1). Nevertheless, weighted estimates often are and should be used because they provide more economic methods for controlling disturbing variables than eliminations of cases.

A numerical example may help explain the increases in variance caused by eliminations and by weighting, as well as the great difference between them. Suppose that for each of five subclasses the sizes of the **a** treatment are equal; thus $n_{ai}/n_a = w_{ai} = 1/5$ for each subclass. Also suppose that the **b** control cases will be eliminated to the **a** standard, though originally the **b** cases were greater in the ratios of $k_i = n_{bi}/n_{ai} = 1, 2, 3, 4$, and 5. Therefore the overall increase of the variance due to elimination of cases in the five subclasses comes to $(1/5)(1 + 2 + 3 + 4 + 5) = 15/5 = 3$. On the other hand, the increase of the variance due to weighting down the n_{bi} cases comes only to 1.37. The following details may be skipped by uninterested readers.

Generally, this increase of the ratio of the variance is $\Sigma w_i k_i$, where k_i is the ratio of elimination and w_i the weight in the ith subclass. If the k_i increase monotonically so that $k_i = 1, 2, 3, \ldots, K$, and the subclass distribution is rectangular so that $w_i = 1/K$, then $\Sigma w_i k_i = (1/K)[K(K + 1)/2] = (K + 1)/2$. This can be large for large values of K (50.5 for $K = 100$) and such a spread in elimination ratios is of first importance. The distribution of weights is of second importance in most situations. Even less important are usually the differences in subclass variances σ_i^2 and costs c_i, which have been neglected here.

For adjustment by weighting we want to compare the variance of the weighted cases with a similar total number $n_b = \Sigma n_{bi}$ of unweighted cases, to be able to judge how much the "effective size" of the sample is reduced by weighting. This and the increase of variance can be measured roughly and quickly by $1 + L = (\Sigma w_i k_i)(\Sigma w_i/k_i)$. Note the new term $(\Sigma w_i/k_i)$, which in our example amounts to $(1/5)(1 + 1/2 + 1/3 + 1/4 + 1/5) = 0.4567$. This new term, $(\Sigma w_i/k_i) \leq 1$, tells us how much lower the variance is for weighting than for elimination by those factors. The variance ratio $1 + L = (\Sigma w_i k_i)(\Sigma w_i/k_i)$ comes to $3 \times 0.4567 = 1.37$, and L stands for the proportionate loss due to weighting. The factor $1 + L$ may be viewed as a "design effect" due to weighting. This is always positive unless the k_i values are all the same constant and cancel leaving $(\Sigma w_i^2) = 1$. The value of 1.37 for $1 + L$ corresponds to $L = 0.37$ for $K = 5$ in row 5 of Table 4.5.1.

The values of $(\Sigma w_i k_i)(\Sigma w_i/k_i)$ are seen there to be sensitive to the spread of k_i, and they increase monotonically to $1 + 1.62 = 2.62$ when $K = 100$. But this is much less than the increase of $(K + 1)/2 = 50.5$ for elimination; less by the ratio $(\Sigma w_i/k_i) = 2.62/50.5 = 0.052$. This results from the decrease of $(\Sigma w_i/k_i)$ as $(\Sigma w_i k_i)$ increases, while the product increases only slowly. In this model we assumed the integral values of $k_i = 1, 2, \ldots, K$. The distribution of weights is of secondary importance and $1 + L$ varies for diverse distributions mostly from 1.2 to 1.4 for $K = 5$; and the 1.37 we saw for the discrete rectangular distribution ($w_i = 1/5$) is higher than most. The highest values of $1 + L$ for any fixed K result from the dichotomy with $W_1 = W_2 = 1/2$; for $K = 5$ this comes to $(1/2)(1 + 5)\cdot(1/2)(1 + 1/5) = 1.8$. For $K = 100$ it goes way up to 25.5 for the dichotomy, but only from 1.3 to 6 for other distributions of the w_i.

Note also that only the n_{bi} would be affected by both elimination and weighting, and the n_{ai} remain unaffected, because all cases were used without weighting. Hence the variance increase of 3 would be affected by elimination in the ratio of $(1 + 3\bar{n}_a/\bar{n}_b)$ up to $K = 5$, and only in the ratio of $(1 + 1.37\bar{n}_a/\bar{n}_b)$ by weighting. These ratios may be much less than 2 if the ratio $\bar{n}_a/\bar{n}_b$ is small, because the pool of n_{bi} cases is much greater than the n_{ai}. The $\bar{n}_a$ and $\bar{n}_b$ denote appropriate averages of the n_{ai} and n_{bi}.

We assumed simple random sampling for variances of σ_i^2/n_i and constant values of σ_i^2, and we disregarded cost factors. All these can be considered in full treatments, which are available with derivations, limitations, and tables elsewhere (Table 4.5.1) [Kish 1974 and 1965a, 11.7].

TABLE 4.5.1. Relative Losses (L) for Six Models of Population Weights (U_i); for Discrete (L_d) and Continuous (L_c) Weights: for Relative Departures (K_i) in the Range from 1 to K[a, b]

Models	K	1.3	1.5	2	3	4	5	10	20	50	100
Dichotomous $U(1-U)$											
(0.5)(0.5)		0.017	0.042	0.125	0.333	0.562	0.800	2.025	4.512	12.005	24.50
(0.2)(0.8)		0.011	0.027	0.080	0.213	0.360	0.512	1.296	2.888	7.683	15.68
(0.1)(0.9)		0.006	0.015	0.045	0.120	0.202	0.288	0.729	1.624	4.322	8.82
Rectangular	L_d	0.017*	0.042*	0.125*	0.222	0.302	0.370	0.611	0.889	1.295	1.620
$U_i \propto 1/K$	L_c	0.006	0.014	0.040	0.099	0.155	0.207	0.407	0.656	1.036	1.349
Linear decrease	L_d	0.017*	0.040*	0.111*	0.203	0.283	0.353	0.616	0.940	1.437	1.917
$U_i \propto K+1-k_i$	L_c	0.006	0.014	0.040	0.097	0.153	0.205	0.409	0.680	1.127	1.514
Hyperbolic decrease	L_d	0.017*	0.040*	0.111*	0.215	0.312	0.404	0.807	1.466	3.014	5.076
$U_i \propto 1/k_i$	L_c	0.006	0.014	0.041	0.103	0.171	0.235	0.528	1.011	2.138	3.621
Quadratic decrease	L_d	0.016*	0.036*	0.080*	0.150	0.211	0.264	0.460	0.696	1.048	1.333
$U_i \propto 1/k_i^2$	L_c	0.006	0.014	0.040	0.099	0.155	0.207	0.407	0.656	1.036	1.349
Linear increase	L_d	0.017*	0.040*	0.111*	0.167	0.200	0.222	0.273	0.302	0.320	0.327
$U_i \propto k_i$	L_c	0.006	0.013	0.037	0.088	0.120	0.148	0.223	0.273	0.308	0.320

[a] From Kish, 1976.
[b] Dichotomous, $1 + L = 1 + U(1-U)(K-1)^2/K$. Also all *. Discrete, $1 + L_d = (\Sigma U_i k_i)(\Sigma U_i/k_i)$, with $k_i = i = 1, 2, 3, \ldots, K$. Continuous, $1 + L_c = \int Uk\, dk \int (U/k) dk$, with $1 \leqslant k \leqslant K$. Only two values, 1 and K, were used for L_d for $K = 1.3$, 1.5, and 2.

4.5B Standardization

For a simple presentation, conforming to that for case-to-case matching, we assumed in Section 4.5A that the W_{ai} will be the standard base to which the w_{bi} become standardized by either elimination or weighting. This describes a common situation in case-by-case matching where the n_{ai} represent a new treatment or a relatively small subpopulation, whereas the n_{bi} represent a larger subpopulation or even the residual population. With standardization, however, one must raise the question of what should be used as its base. We mention here six alternatives from a larger possible number: Choices can vary and they should depend on the substantive needs of research. Assume *relative* standard weights for each alternative, so that $\Sigma W_{si} = 1$, and $W_{si} = n_{si}/\Sigma n_{si}$. The adopted base W_{si} may be any of the following.

1. The weights of W_{ai} for the **a** subpopulation, representing a new, changed, or experimental treatment; or representing an unusual or rare subclass.
2. The weights W_{bi} for the b subpopulation, representing the standard, old, accepted, or "control" treatment; or representing the residual or the entire population.
3. The arithmetic mean $(W_{ai} + W_{bi})/2$ of (1) and (2).
4. The geometric mean $\sqrt{(W_{ai}W_{bi})}/\Sigma\sqrt{(W_{ai}W_{bi})}$ or $\sqrt{(n_{ai}n_{bi})}/\Sigma\sqrt{(n_{ai}n_{bi})}$, or some other average.
5. Weights W_i from some standardized population, other than either a or b, are often used in demography and economics.
6. Statistical theory provides a clear answer, although to a somewhat different question: how to choose the W_i to minimize the variance of $\Sigma W_i(\bar{y}_{ai} - \bar{y}_{bi})$. This differs from the substantive questions addressed by alternatives 1 to 5, but it may be particularly appropriate for experimental situations with small n_{ai} and n_{bi}. Then the W_i should be chosen to be inversely proportional to their variances: Make $W_i \propto 1/(\sigma^2_{ai}/n_{ai} + \sigma^2_{bi}/n_{bi})$. When the unit variances are (approximately) equal, make $W_i \propto n_{ai}n_{bi}/(n_{ai} + n_{bi})$ [Kalton 1968; Keyfitz 1953]. These assume simple random sampling and they disregard cost factors, which can be introduced along with design effects [Kish 1976], (7.1).

Fortunately, whether we use one set of weights or another usually (though not always, nor necessarily) makes only for smaller (second-order) differences, compared with frequently large (first-order) differences resulting from merely using common weights to control for the effects of disturbing variables. To demonstrate this, let us use a simple model to decompose the effects of weighting on the difference of two means:

$$\begin{aligned}
\bar{y}_a - \bar{y}_b &= \Sigma\, w_{ai}\bar{y}_{ai} - \Sigma\, w_{bi}\bar{y}_{bi} \\
&= \Sigma\, w_{ai}(\bar{y}_{ai} - \bar{y}_{bi}) + \Sigma\, \bar{y}_{ai}(w_{ai} - w_{bi}) - \Sigma\,(w_{ai} - w_{bi})(\bar{y}_{ai} - \bar{y}_{bi}) \\
&= \Sigma\, w_{bi}(\bar{y}_{ai} - \bar{y}_{bi}) + \Sigma\, \bar{y}_{bi}(w_{ai} - w_{bi}) + \Sigma\,(w_{ai} - w_{bi})(\bar{y}_{ai} - \bar{y}_{bi}) \\
&= \Sigma\, w_{bi}(\bar{y}_{ai} - \bar{y}_{bi}) + \Sigma\, \bar{y}_{ai}(w_{ai} - w_{bi}).
\end{aligned} \tag{4.5B1}$$

View the first term as the component for the difference of means that interests us, freed from the disturbing effects of differences $(w_{ai} - w_{bi})$ in the weights of subclasses. The second term is the component due to the weight differences to be removed; and it is removed when we use the first component, but only to the degree that we can neglect the third component. This third component is an "interaction" term that is small to the degree that weight differences are not consistently and strongly correlated with the differences of the subclass weights. Another view of this component is also instructive:

$$\Sigma\, w_{ai}(\bar{y}_{ai} - \bar{y}_{bi}) - \Sigma\, w_{bi}(\bar{y}_{ai} - \bar{y}_{bi}) = \Sigma\,(w_{ai} - w_{bi})(\bar{y}_{ai} - \bar{y}_{bi}). \tag{4.5B2}$$

Thus when the "interaction" component is negligible it matters little whether we use the w_{ai}, the w_{bi}, or some average value between them. In most situations of negligible interaction the same probably holds for reasonable bases from other sources; these must be reasonable bases because mathematical contradictions can always be found.

We shall explore components further in Section 4.5C on indexes, and we merely note here that the choice of weights can be important in some situations. This is clear to economists concerned with price indexes, who spend much effort on the choice of weights for different items to measure the movement of prices. For example, for decades in the United States the prices of electrical goods remained steady, for electronic goods they decreased, while medical expenses "skyrocketed." International comparisons are most revealing: U.S. newspapers often display the high prices of autos, TV sets, clothes, and blue jeans in Communist countries. But they rarely report that their rents are only 5 to 10 percent of income; that health care, education, books, amusements, and public transport are almost free; and that basic foods sell (often rationed) for low prices (subsidized from the high prices of autos, TV, clothes, etc.). Without expressing preferences, we comment on the biased weighting of prices in the media—ours and probably theirs. The message is all in the weights, and they do matter there.

Large differences due to the choice of a standard base can therefore occur in diverse social sciences. However, different *reasonable* bases for standards will usually yield similar results. Perhaps the easiest check in many situations is to compute the $(\bar{y}_{ai} - \bar{y}_{bi})$ differences with both w_{ai} and w_{bi} for weights. As (4.5B2) shows, a negligible difference between them indicates small "interaction." If not negligible, it may provide clues for directions to further research for better standards. Computing and presenting the differences with different standards may also be helpful and stimulating to the consumers of data. Literature on the justification, theory, and methods for standardization exists in economics, demography, and health [Shryock and Siegel 1973, pp. 418–423, 481–486; Hill 1961, Ch. 17; Yule and Kendall 1950, Ch. 25; Mueller et al. 1970, Sec. 7.2].

Most literature on standardization, like our presentation, uses finite subclasses. However, continuous models can also be used for the "handling of disturbing variables" [Cochran 1965, Sec. 3; Cochran 1983, Ch. 6]. Some interesting theoretical results can be facilitated with strong, continuous models. However, empirical data may be better handled with finite subclasses. Furthermore, continuous linear regression models need stronger and different assumptions than most social research can support.

4.5C Indexes

A small change of symbols will simplify this presentation and relate it more closely to the standard literature on indexes: Denote the treatment with small case and the standard base with capitals. The bases often come either from censuses or from models and hence are free of the sampling variation that affects the treatment data. Thus:

$$\text{treatment } \bar{y} = \Sigma\, w_i y_i \qquad \text{and} \qquad \text{BASE } \bar{Y} = \Sigma\, W_i Y_i.$$

Set the weights $\Sigma\, w_i = \Sigma\, W_i = 1$; if other kinds of weights are actually used they should be divided ("normalized") by $\Sigma\, w_i$ or $\Sigma\, W_i$. The indicator ($i = 1, 2, \ldots, I$) can be, and often is, based on multivariate subclassifications. But to use separate indicators for each category would be unnecessary and confusing. The y_i and Y_i usually denote subclass means (for treatment and base) but we have not used the bars over them.

Direct standards and indirect standards are both commonly used. In the

$$\text{direct standard} = \bar{y}_d = \Sigma\, W_i y_i, \tag{4.5.C1}$$

the subclass means are standardized by multiplication with base weights. The

$$\text{indirect standard} = \bar{y}_{nd} = \bar{y}\frac{\bar{Y}}{\Sigma\, w_i Y_i} = \Sigma\, w_i y_i \frac{\Sigma\, W_i Y_i}{\Sigma\, w_i Y_i} \tag{4.5.C2}$$

needs to be used when the y_i for the subclasses are not available, would need too many computations, or would be too unstable. But the base means Y_i

and the treatment weights w_i are available and stable, as are the overall means $\bar{y}$ and $\bar{Y}$ for both treatment and base.

From either standard several indexes can be computed, as they often are for economic, demographic, health, and other indicators. They are most commonly used for time series of values that are standardized to some base period; an average of several periods (years) may provide more stable bases. But some indexes may also refer to spatial bases, such as national bases for regional or local data. The names for these indexes vary, with frequent use of personal names that I shall mostly avoid. Commonly, the indexes are multiplied by 100 to express percentlike fluctuations. The

$$\text{average of relatives} = \Sigma W_i \frac{y_i}{Y_i} \qquad (4.5.C3)$$

seems to be the simplest and most direct use of the ratios y_i/Y_i of treatment to base, with base sizes W_i of subclasses for weights. But other indexes are usually preferred because of its disadvantages: (1) the values of y_i must be computed and (2) some y_i may be small and unstable; (3) when the Y_i are not constants (known or fixed) but random variables, the fluctuations of the denominators cause instability, and a technical bias. This resembles the problems of "separate ratio estimators" (4.7).

It would also be possible to use $\Sigma(w_i y_i/Y_i)$, if the $(w_i y_i)$ were known but not the y_i (or the w_i) separately. Also the W_i could conceivably be the same constant $1/I$. Or the weights could be W_{0i} from a base other than the Y_i. Such problems are avoided in the

$$\text{aggregative index} = \frac{\Sigma W_i y_i}{\Sigma W_i Y_i} = \frac{\bar{y}_d}{\bar{Y}} = \frac{\text{direct standard}}{\text{base}} = \Sigma \frac{W_i Y_i}{\Sigma W_i Y_i} \cdot \frac{y_i}{Y_i}. \qquad (4.5.C4)$$

Note that in this "combined ratio estimator" we have a ratio of sums rather than the sum of individual ratios in (4.5.C3), hence more stability. These are favored in survey sampling and their properties have been much studied (4.7). From the last expression note that, compared with (4.5.C3), the relative weights W_i are changed to be proportional to $W_i Y_i$, which may also make substantive sense for many indexes. Still in the choice of the sizes to be used for the W_i, flexibility permits, and economists employ, several kinds of aggregative indexes. Conforming to common economic usage, we shall use W_{0i} and Y_{0i} to denote data from a base year (0). With $(W_i = w_{0i})$ the

$$\text{base standard} = \frac{\Sigma W_{0i}\, y_i}{\Sigma W_{0i} Y_{0i}} = \frac{\Sigma W_{0i} y_i}{\bar{Y}_0} = \frac{\text{direct standard}}{\text{base mean}}, \qquad (4.5.C5)$$

also known as Laspeyres' index. With $(W_i = w_i)$ the

$$\text{current standard} = \frac{\Sigma\, w_i y_i}{\Sigma\, w_i Y_{0i}} = \frac{\bar{y}}{\Sigma\, w_i Y_{0i}} \left(\frac{\bar{Y}}{\bar{\bar{Y}}}\right) = \frac{\text{indirect standard}}{\text{base}}, \tag{4.5.C6}$$

also known as Paasche's index, which does not need the separate values of y_i, similar to the indirect standard. With $[W_i = (W_{0i} + w_i)/2]$ the

$$\text{average standard} = \frac{\Sigma\,(W_{0i} + w_i) y_i}{\Sigma\,(W_{0i} + w_i) Y_i} = \frac{\bar{y} + \text{direct}}{\bar{Y} + \Sigma\, w_{0i} Y_i} \tag{4.5.C7}$$

is based on the arithmetic mean of W_{0i} and w_i. It equals $\bar{Y}(1 + \bar{y}/\text{indirect})$.

$$\text{Fisher's "ideal" index} = \sqrt{\left(\frac{\Sigma\, W_{0i} y_i}{\Sigma\, W_{0i} Y_{0i}} \cdot \frac{\Sigma\, w_i y_i}{\Sigma\, w_i Y_{0i}}\right)} = \sqrt{\frac{(\text{direct} \times \text{indirect})}{\text{base}}}$$

$$= \sqrt{\left(\frac{\bar{y}}{\bar{Y}} \cdot \frac{\Sigma\, W_{0i} y_i}{\Sigma\, w_i Y_{0i}}\right)}$$

$$= \sqrt{(\text{base} \times \text{current standard})}, \tag{4.5.C8}$$

and is based on the geometric mean of the base and the current standards, as a compromise between them. The simple aggregative index uses equal weights $W_i = 1/I$:

$$\text{simple aggregative index} = \frac{\Sigma\, y_i}{\Sigma\, Y_{0i}} \quad \text{or} \quad \frac{\Sigma\, u_i y_i}{\Sigma\, u_i Y_{0i}}, \tag{4.5.C9}$$

where u_i = number of units in the interval, where these can differ in length.

Proportions (percentages) comprise a common type of variable, where $y_i = p_{ik} = n_{ik}/n_i$, the proportion of the kth category in the ith subclass. When the weight in the base is $W_i = N_i/N$, the

$$\text{direct standard} = p_k(d) = \Sigma\, W_i p_{ik} = \Sigma\,(N_i/N) n_{ik}/n_i. \tag{4.5.C10}$$

When the n_{ik} are not available, one may use the

$$\text{indirect standard} = p_k(nd) = \frac{p_k P_k}{\Sigma\, w_i P_{ik}}, \quad \text{with} \quad P_{ik} = \frac{N_{ik}}{N_i}. \tag{4.5.C11}$$

These would do well enough for single categories (k), but often when the sum over all categories is needed, it turns out that the values for $p_k(nd)$ do not sum to 1.00 (or 100 percent), as one would wish. Therefore some [Kitagawa 1964] recommend using instead $P_k + p_k - \Sigma w_i P_{ik} = n_i/n + N_i/N - \Sigma (n_i/n)(N_i/N)$.

Comparisons of indexes are sometimes needed, e.g., for differences of periods. A brief look at their structures in their simplest forms may be useful for understanding the nature of their sampling variances [Kish 1968]. Comparison with the base which has been normalized to unity (or 100 percent), $\bar{Y} = \Sigma W_i Y_i = 1$ can be written for the average of relatives as

$$\frac{\Sigma W_i y_i}{Y_i} - 1 = \frac{\Sigma W_i y_i}{Y_i} - \Sigma W_i Y_i = \frac{\Sigma W_i(y_i - Y_i)}{Y_i},$$

and for aggregatives as

$$\frac{\Sigma W_i y_i}{\bar{Y}} - 1 = \frac{\Sigma W_i(y_i - Y_i)}{\Sigma W_i Y_i} = \Sigma \frac{W_i Y_i}{\Sigma W_i Y_i} \frac{(y_i - Y_i)}{Y_i}.$$

The difference of aggregatives for two periods will be $\Sigma W_i(y_i - y_i')/\Sigma W_i Y_i$.

4.5D Components of the Differences $(y - y')$ and $(Y - Y')$

Separation into components of differences between means has been an empirical tool especially in demographic research since three publications [Kitagawa 1955, 1964; Althauser and Wigler 1972]. These papers, and others since then, give more detailed descriptions and examples of applications. Whereas standardization aims chiefly at removing disturbing variables, component analysis is a research tool for separating and evaluating two or more determinants (predictors) of the predictand variables y. The concept and algebra are similar to those of (4.5B), but the aims here differ somewhat and include additional development. The notation is somewhat simpler here than in either (4.5B) or in Kitagawa's papers.

$$\begin{aligned} \bar{y} - \bar{Y} &= \Sigma w_i y_i - \Sigma W_i Y_i = \Sigma W_i(y_i - Y_i) + \Sigma y_i(w_i - W_i) \\ &= \Sigma W_i(y_i - Y_i) + \Sigma Y_i(w_i - W_i) + \Sigma (y_i - Y_i)(w_i - Y_i) \\ &= \Sigma W_i dy_i + \Sigma Y_i dw_i + \Sigma dy_i dw_i. \end{aligned} \qquad (4.5.D1)$$

The first line has only two terms but is less meaningful because of changes in two sources of multipliers. The three terms on the following lines are the components, respectively, for rates (or variables), for weights (or composition), and for interaction. Kitagawa's exposition for rates is devoted to proportions, but extensions to other variables can be directly made. The W_i and Y_i refer to standard bases, and for differences between two standardized differences we have:

$$\begin{aligned}(\bar{y} - \bar{y}') &= \Sigma W_i(dy_i - dy_i') + \Sigma Y_i(dw_i - dw_i') + \Sigma (dy_i dw_i - dy_i' dw_i') \\ &= \Sigma w_i(y_i - y_i') + \Sigma y_i(w_i - w_i') - \Sigma (w_i - w_i')(y_i - y_i').\end{aligned} \tag{4.5.D2}$$

We note (as in Section 4.5B) that $\Sigma (y_i - y_i')(w_i - w_i') = \Sigma w_i(y_i - y_i') - \Sigma w_i(y_i - y_i')$, so that the interaction term equals the effect of different weights on the difference of rates, which is often relatively small—though not always. The interaction also disappears with the mean of the two multipliers:

$$\bar{y} - \bar{y}' = \Sigma 0.5(w_i + w_i')(y_i - y_i') + \Sigma 0.5(y_i + y_i')(w_i - w_i').$$

Furthermore, the second term for the difference in weights can be decomposed into components for the diverse sources of the effects I and J:

$$\begin{aligned}&\Sigma\Sigma\, w_{ij}y_{ij} - \Sigma\Sigma\, w_{ij}'y_{ij}' \\ &= \Sigma\Sigma\, 0.5(w_{ij} + w_{ij}')(y_{ij} - y_{ij}) \quad \text{difference of rates} \\ &+ \Sigma_j 0.5(w_j + w_j')\, \Sigma_i 0.5(y_{ij} + y_{ij}')\left(\frac{w_{ij}}{w_j} - \frac{w_{ij}'}{w_j'}\right) \quad I \text{ component} \\ &+ \Sigma_i 0.5(w_j + w_j')\, \Sigma_j 0.5(y_{ij} + y_{ij}')\left(\frac{w_{ij}}{w_i} - \frac{w_{ij}'}{w_i'}\right) \quad J \text{ component} \\ &+ \Sigma_i\Sigma_j\, 0.5(y_{ij} + y_{ij}')\left[\left(\frac{w_{ij}'w_i}{w_i'} - \frac{w_{ij}w_i'}{w_i}\right) + \left(\frac{w_y'w_j}{w_j'} - \frac{w_{ij}w_j'}{w_j}\right)\right] \\ &\qquad IJ \text{ component.}\end{aligned} \tag{4.5.D3}$$

4.6 COVARIANCES AND RESIDUALS FROM LINEAR REGRESSIONS; CATEGORICAL DATA ANALYSES

4.6A Covariance Analysis

Covariance analysis denotes a method for removing the effects of disturbing variable(s) **x** from differences $(\bar{y}_a - \bar{y}_b)$ by computing the covariance of **x** with **y** and adjusting for its regression effects. (I use D for disturbing variables elsewhere, but x fits more of the references.) The method can be somewhat more efficient than standardization with few subclasses; but we shall see later that three, four, or five subclasses can perform quite well. Covariance analysis is also more clearly and elegantly based on explict models; and its theory is prominent in many statistics books [Anderson and Bancroft 1952, Ch. 21; Cochran and Cox 1950, 3.8; Kirk 1968, Ch. 12; Snedecor and Cochran 1967, Ch. 4]. I quote below from "Analysis of covariance; its nature and uses" [Cochran 1957] in the issue of *Biometrics* devoted to articles on covariance.

The method, however, is not applied often in empirical and social research because of the following severe limitations.

1. Its linear model postulates that the effects of treatment, blocking, and regression be additive.
2. The model is based on continuous data; this is severe obstacle for most social research, which is based on categorical data with nominal or ordinal scales. Several new methods for analyzing categorical data have been developed recently to overcome these limitations (4.6C).
3. The residuals e_{ij} are postulated to have zero means and normal, independent, and identical distributions (the same variances). These assumptions are needed for tests of significance in classical presentations, but some robust methods developed in survey sampling for errors may be applied here also (7.1).
4. The simple, common format is designed for linear regression on a single disturbing variable, but we may desire to remove simultaneously the effects of several disturbing variables. For such multiple regressions more complex methods have been developed.
5. Analysis and computations can be fairly complex, especially for the multivariate case, and modern computers are making these more feasible. However, computing packages can also distance researchers from the meaning of the final statistical output and from the ends of statistical inference.
6. Interpretation of the results of covariance analyses can become subtle and doubtful—as controversies in the literature show [e.g., Cochran 1969].
7. The usual classical presentation of covariance analysis is framed in terms of tests of significance, but the aims of social and other research are more often the magnitudes of values and relations (1.5).

Most of these limitations can be avoided—partly or entirely, readily or carefully—with methods based on subclasses for controlling disturbing variables. Nevertheless, a brief look at the covariance model is worthwhile, both because it can be used occasionally and because its model is broadly instructive.

> The typical mathematical model appropriate to the analysis of covariance ... from the viewpoint of covariance ... may be rewritten as
>
> $$y_{ij} - \beta(x_{ij} - x_{..}) = \mu + \tau_i + \rho_j + e_{ij}. \qquad (4.6.1)$$
>
> Here $\mathbf{y}_{ij}$ is the yield or response, while $\mathbf{x}_{ij}$ is an auxiliary variate, sometimes called the concomitant variate or *covariate*, on which $\mathbf{y}_{ij}$ has a linear regression with regression coefficient β. The constants, μ, τ_i and ρ_i are the true mean and the effects of the ith treatment and jth replication, respectively. The residuals $\mathbf{e}_{ij}$ are random variates, assumed in standard theory to be normally and independently distributed with mean zero and common variance.... The symbols $x_{..}$, $y_{..}$ denote overall means.... The quantities $y_{ij} - \beta(x_{ij} - x_{..})$ are the deviations of y_{ij} from its linear regression on $\mathbf{x}_{ij}$, or the values $\mathbf{y}_{ij}$ after adjustment for this linear regression. In this setting, τ_i may be regarded as the true effect of the ith treatment on $\mathbf{y}_{ij}$, after adjustment for the linear regression on the covariate $\mathbf{x}_{ij}$. Thus the technique enables us to remove that part of an observed treatment effect which can be attributed to a linear association with the $\mathbf{x}_{ij}$ ... the analysis of covariance extends the study of regression relationships to data of complex structure in which the nature of the regression is at first obscured by structural effects like τ_i and ρ_i. [Cochran 1957]

Cochran also lists four other uses of covariance analysis: to increase precision in randomized experiments, to throw light on the nature of treatment effects, to fit regressions in multiple classifications, and to analyze data when some observations are missing. Here we are interested in covariance chiefly to remove the biasing effects of disturbing variables in observational studies, but it also increases their precision as it does in randomized experiments. The difference $(y_a - y_b)$ of means between two groups is

$$(\bar{y}_a - \bar{y}_b) = \tau_a - \tau_b + \beta(x_a - x_b) + e_a - e_b. \qquad (4.6.2)$$

The unadjusted difference thus includes not only the difference $(\tau_a - \tau_b)$ of true means and the error term $e_a - e_b$, but also the bias term $\beta(x_a - x_b)$. But this term is removed by the covariance adjustment. However,

> we can never be sure that bias may not be present from some disturbing variable that was overlooked. In randomized experiments, the effects of this variable are distributed among the groups by the randomization in a way that is taken into account in the standard tests of significance. There is no such safeguard in the absence of randomization. [Cochran 1957]

TABLE 4.6.1. Proportion of Constant Bias Removed with Different Numbers of Subclasses

Numbers of subclasses	1	2	3	4	5	6	∞
Proportion of bias removed	0	.64	.80	.86	.90	.92	1.00

For all methods of controlling disturbing variables, questions arise about their adequacy in the face of errors of measurement. One problem concerns the loss of control due to grouping into few subclasses an underlying variable that is continuous, or is approximately continuous with many potential classes. This common problem can arise because either the variable gets measured that way or it is reduced later to a few subclasses. This problem receives its best treatment from Cochran [1968] for a wide range of distributions of the disturbing variable. His results have broad meaning, with reassuring uniformity of consequences for different distributions of the disturbing variable, and Table 4.6.1 presents a basic summary. It shows that, e.g., five control classes will reduce the relative bias of a continuous disturbing variable by a factor of 0.90; for example, from 8 percent to 0.8 percent.

Further, Cochran also deals with random error, which introduces misclassification into the subclasses. Such misclassifications reduce the preceding ratio of effectiveness by the factor $(1 + h)$, where h is the additional variance due to random errors of classification; this is computed as a ratio to the variance caused by an errorless disturbing variable in the predictor.

Table 4.6.2 presents an example of the use of controls to remove the effects of a single disturbing variable, age. At first glance the unadjusted cigarette smokers seem to do as well as nonsmokers, whereas cigar and pipe smokers appear to live more dangerously. The differences in mean ages, though small, raises doubts because age is strongly related to death rates and also to cigarette, pipe and cigar smoking. Adjusting for age subclasses (of smokers to nonsmokers) brings out step by step a corrected relationship of smoking habits to death rates: Cigarettes are deadly; cigars and pipes are not. Nonsmokers' rates remain at 13.5 because they serve as the base to which the others are adjusted. We may guess that 11 subclasses are almost as good as a continuous adjustment, and 5 subclasses would be close to them. Canadian and British data lead to similar conclusions [Cochran 1968].

Finally, this example also serves to make another point, not strange to empirical researchers, about controls for disturbing variables. Commonly, such controls serve to probe the validity of differences found in the predictand variables, and imposing controls often reduces those differences. In this case, however, nonsmokers and cigarette smokers show no differences

TABLE 4.6.2. Adjusted Death Rates per 1000 U.S. Adult Males, Using 2, 3, and 11 Subclasses Compared to Unadjusted (1) Rates[a]

	Mean Age	Number of Subclasses: 1: Unadjusted	2	3	11
Nonsmokers	57.0	13.5	13.5	13.5	13.5
Cigarettes only	53.2	13.5	16.4	17.7	21.2
Cigars and pipes	59.7	17.4	14.9	14.2	13.7

[a]From Cochran 1968.

at first, but imposing proper controls uncovers the differences, which were hidden by the disturbing variable of age subclasses with contradicting death rates. After adjusting for the bias of their youth, cigarette smokers are shown to be in mortal danger.

4.6B Residuals

We may begin with a regression model $y_j = \Sigma b_p x_p + e_j$, where $\Sigma b_p x_p = b_0 + b_1 x_1 + b_2 x_2 + \cdots$ expresses a multivariate linear regression model, and e_j is the error term for the jth observation y_j. This differs from (4.6.1): it does not include a population mean τ_i and is cast in a model for regression rather than for classical analysis of variance. The principal difference concerns the role for e_j, which in (4.6.1) was merely an error term assumed to be normal with zero mean, independently and identically distributed, and hopefully small. Here $e_j = y_j - \hat{y}_j$ becomes the object of attention, where $\hat{y}_j = \Sigma b_p x_p$; and

> the residuals e_j are the differences between what is actually observed, and what is predicted by the regression equation—that is, the amount which the regression equation has not been able to *explain*.... We now give ways of examining the residuals in order to check the model. These are graphical, are easy to do, and are usually very revealing when the assumptions are violated. The principal ways of plotting the residuals are 1. Overall; 2. In time sequence, if the order is known; 3. Against the fitted values; 4. Against the independent variables; 5. In any way that is sensible for the particular problem under consideration. [Draper and Smith 1966, Ch. 3]

These examinations of residuals (or deviations) by statisticians [e.g., Anscombe and Tukey 1963; Neter and Wasserman 1974, Secs. 4.2, 13.5, 15.4] begin with the aim "to check the model." If the linear regression represents the predictor variables, the residuals may represent the disturbing variables to be examined and analyzed. It may begin with a graphical, optical examination, but also continue with other methods such as inclusion in the regression model, or analysis of subclasses and/or of variances, etc.

However, with a drastic reversal of roles (we suggest), the linear regression can represent the disturbing variables, as with covariance methods. But now instead of trying to explain and relate the variable y_j to other variables, the analyses proceed with the residual $e_j = y_j - \hat{y}_j$. Analysts of those relations combine substantive and statistical aspects and can include analyses of subclasses, of variances, of regressions, etc.

Further extensions of residual analyses are also possible.

1. Other models of regression may be used, such as nonlinear and categorical data analysis (4.6C).
2. Other theoretical models may be used instead of regressions.
3. Instead of differences, the relations of actual to predicted may be represented by ratios or other functions.

The last two points are well represented by an interesting example from Coleman [1964, Ch. 15]:

> . . . in considering a given complex social phenomenon, certain aspects of it are explainable by "sociologically trivial" assumptions, or by matters irrelevant to the substantive matters under investigation . . . the approach suggested here is to make some simple and reasonable null assumptions, and then to use the *deviations* from predictions consequent upon the assumptions as a measure of various matters, as, for example, the "social distance" between two groups.

Then in his example of travel and communication between pairs of cities

> . . . under these null or "ideal" conditions, the rates of interaction between two areas which contain n_1 and n_2 persons and which are d_{12} distance apart will be proportional to $n_1 n_2 / d_{12}$. . . . This factor, $n_1 n_2 / d_{12}$, is the same one which has been proposed as "the law of social gravity." . . . In contrast, what is proposed here is that this factor be used simply as a base line, or standardization, . . .

From data on numbers x_{12} of airline trips between pairs of cities, travel rates are computed, standardized both for populations n_1 and n_2 and for distances d_{12} between them: then $w_{12} = x_{12} d_{12} / n_1 n_2$. The mean of these for $N(N - 1)$ pairs between N cities is $\bar{w} = \Sigma\Sigma w_{12} / N(N - 1)$. Then $Z_{12} = w_{12} / \bar{w}$ shows deviations above and below 1 for the average. A triangular matrix of these $N(N - 1)/2$ pairs shows some regularities, most due probably to some cities (e.g., Washington) showing generally larger values than 1 and others (e.g., Philadelphia) showing generally less than 1. These average values may be factored out in $V_{12} = Z_{12}/Z_1Z_2$. These values exhibit individual relative deviations specific to each pair.

4.6C Nonlinear Regressions, Dummy Variables, MCA, Categorical Data Analysis

This will have to be a brief, superficial view of a broad and varied field of statistical analysis, and further restricted to the use of regression for controlling disturbing variables, rather than for principal analysis. First, let us look at several kinds of departures from linear continuous models to see whether they can be treated with the covariance or the residuals of (4.6A) or (4.6B) or left for the subclass treatments of (4.1) to (4.5), or subjected to one of the methods described here.

1. *Monotonic relations, whether continuous or ordinal,* can sometimes be handled well enough by linear regression. Transformations, with a mathematical model guided by substantive theory, may give better results—but seldom with small samples subject to their large measurement errors and to the small explanatory powers of the usual data of social research. Results from several transformations may be compared with each other and with the "safer" results of simple subclass analysis.
2. *Curvilinear, nonmonotonic, continuous* relations may be poorly estimated by linear regressions. Either transformations or the higher order terms usually recommended for "curvilinear" regressions may be better, with models skillfully guided by theory.
3. *Categorical predictors, with predictands either continuous or dichotomous,* are often treated with "dummy variable" regressions or MCA programs.
4. *Categorical data analysis* can be applied to all data that are either intrinsically categorical or deliberately categorized for analysis.

Dummy variable analysis has mathematical expressions that are similar to the regressions equation:

$$y_i = \Sigma b_i x_i + e_j = b_0 + b_1x_1 + b_2x_2 + b_3x_3 + b_4x_4 + \ldots + e_j. \tag{4.6.3}$$

This additive expression may have many terms, which fast computers can handle, and modern literature has programs and descriptions for them

[Draper and Smith 1966, Sec. 5.3; Suits 1957]. For example, x_1 may represent the two categories of a dichotomy, with $x_1 = 1$ for the first category and $x_1 = 0$ for the second. A trichotomy could be represented by x_2 and x_3 with (1, 0), (0, 1) and (0, 0) for the three categories. A polytomy with k categories needs $(k - 1)$ terms in the equation. A continuous (or monotonic) variable may be represented by x_4. The predicted value is $\hat{y}_j = \Sigma b_i x_i$ and the literature discusses the meaning of the regression coefficients b_i. But here we regard it as a method for removing disturbing variables in order to examine the adjusted values $e_j = y_j - \hat{y}_j = y_j - \Sigma b_i x_i$ and to search for relations among other explanatory variables.

Multiple classification analysis (MCA) is a computer program for the analysis of dummy variables for multiple regression with categorical predictors [Andrews et al. 1973]. MCA

> is a technique for examining the interrelationships between several predictor variables and dependent variables within the context of an additive model. Unlike simpler forms of other multivariate methods, the technique can handle predictors with no better than nominal measurement, and interrelationships of any form among predictors or between a predictor and the dependent variable. The dependent variable, however, should be an intervally scaled (or a numerical) variable without extreme skewness, or a dichotomous variable with two frequencies which are not extremely unequal. The statistics printed by the program show how each predictor relates to the dependent variable, both before and after adjusting for the effects of other predictors, and how all the predictors considered together relate to the dependent variable.

(See also descriptions in Morgan et al. 1962, App. E, and in Duncan and Blau 1967, pp. 128–140.) MCA has been applied often to data adjusted for disturbing variables.

In Table 4.6.3 the deviations in the last column have been adjusted for stage in life cycle (.39), income (.38), physical condition (.16), region (.06), state old age assistance (.06), color (.04), and sex (.01); beta coefficients are given in parentheses. Note that for under age 25 the proportion living with relatives drops from .17 + .44 = .61 to .17 + .15 = .32 after adjustment. Also for the old (75 and over) the proportion drops from 29 percent to 16 percent after adjustment. Thus we see drastic changes in the relationship due to adjustments, especially for life cycle and income. This book is full of such tables, though most of the results of adjustments are not as dramatic as this example for age. For example, for those with less than $1000 income, the proportion living at home goes from 70 to 52 percent after adjustment.

Categorical data analysis has enjoyed remarkable development recently in various forms and especially with loglinear models. Its most humble and

TABLE 4.6.3. Living in Relatives' Home: Deviations for Age of Heads from Grand Mean of .17[a]

Age of Heads	Number of Adult Units	Percent of Adult Units	Unadjusted Deviations[b]	Adjusted Deviations[b]
Under 25	434	12.5	.44	.15
25–34	631	18.6	−.03	.03
35–44	694	20.6	−.11	−.03
45–54	624	17.9	−.11	−.03
55–64	517	14.5	−.09	−.05
65–74	317	10.2	−.01	−.04
75 or older	179	5.7	.12	−.01

[a]From Morgan et al. 1962, Table 14.6.
[b]Deviations from grand mean of .17.

most common forms are the well-known 2 × 2 tables for double dichotomies. However, with modern computers multidimensional polytomies with count data in $p \times q \times r \times s$ cells can be handled. The large literature, unfortunately, is still mostly devoted to tests of significance, but methods for estimates are also emerging. However, I know of no applications for controlling disturbing variables to yield adjusted explanatory variables. But perhaps these will also emerge in the future. [Agresti 1984].

4.7 RATIO ESTIMATES

In this section I transgress (reluctantly) my confinement to design problems and venture briefly into a special area of estimation. This venture seems advisable because of the book's many references to ratio estimates and because of their fundamental position in survey sampling. This topic is well covered and developed at length and in breadth in all textbooks on survey sampling [Hansen, Hurwitz, and Madow 1953, 4.16–4.21; Cochran 1977, Ch. 6; Kish 1965, Ch. 6]. However, I felt it advisable to include this brief outline for those not familiar with this topic from sampling texts. In complex samples means have the form $r = y/x$ because the sample size is not a fixed n but a variable x, subject to several sources of variation. The topic also belongs to this chapter on controls because the ratio mean r is also used often, as discussed later, for adjusted means like Kr and like $\Sigma W_h r_h$, where K and W_h are factors known from outside sources.

It is best to plunge directly into the stratified clustered selections that characterize survey sampling. *Primary selections* are made separately from each of H strata ($h = 1, 2, \ldots, H$), and the number of these *primary selections* from the hth stratum is denoted as $a_h(\alpha = 1, 2, \ldots, a_h)$. A common and useful design uses *paired selections* of two units from each stratum when $a_h = 2$ and ($\alpha = 1, 2$). For the αth primary selection the sums for variables Y and X may be denoted as $y_{h\alpha}$ and $x_{h\alpha}$ and the ratio mean as:

$$r = \frac{y}{x} = \frac{\Sigma_h y_h}{\Sigma_h x_h} = \frac{\Sigma_h \Sigma_\alpha y_{h\alpha}}{\Sigma_h \Sigma_\alpha x_{h\alpha}} \qquad (4.7.1)$$

$$= \frac{\Sigma_h (y_{h1} + y_{h2})}{\Sigma_h (x_{h1} + x_{h2})}, \quad \text{for paired selections} \qquad (4.7.1')$$

$$= \frac{\Sigma_j y_j}{\Sigma_j x_j}, \quad \text{the mean for all cases.} \qquad (4.7.2)$$

The importance and generality of this ratio mean may be clarified with several remarks.

1. This ratio mean is the most common and general statistic based on complex (clustered or multistage) samples in actual survey research.
2. The last form (4.7.2) shows that this represents the simple mean of all cases without regard to strata and selection units. But using $\bar{y} = \Sigma y_j/n_f$ (with fixed size n_f) would disregard the forms that the variance computations must use. The stratified clustered form (4.7.1) displays the terms used in variance computations.
3. It is most useful that the terms $y_{h\alpha}$ and $x_{h\alpha}$ are adequate for variance computations with the primary selections (or ultimate clusters) model regardless of other components for later stages and strata for selections (7.1.E).
4. Those terms are the sums for primary selections of the Y and X variables for individuals cases: $y_{h\alpha} = \Sigma_j y_{h\alpha j}$ and $x_{h\alpha} = \Sigma_j x_{h\alpha j}$. But the denominator is most often simply a case count so that $x = n$, $x_{h\alpha} = n_{h\alpha}$, and $x_{h\alpha j} = 1$, a count variable. For weighted means $x_{h\alpha} = n_{h\alpha} = \Sigma_j w_{h\alpha j}$ and $y_{h\alpha} = \Sigma_j w_{h\alpha j} y_{h\alpha j}$. Also r denotes a proportion for a dichotomy when $y_{h\alpha j} = 0$ or 1, among the $x_{h\alpha}$ case counts.
5. The $x_{h\alpha}$ and the $x = \Sigma_h \Sigma_\alpha x_{h\alpha}$ are variable even when they denote case counts, because sample sizes in the sampling units are typically not fixed, but variable. In addition to ignorance of cluster sizes, there are

also variations due to nonresponses. But the greatest variations typically occur for subclasses that are crossclasses. Nevertheless, the ratio mean can deal effectively with all those sources of variation.

6. However, it is clear that, despite those variations, the denominator x should not be allowed to approach zero because the ratio y/x would become unstable. It is good practice, based on theory, that the coefficient of variation of x, C_x, be checked to be less than 0.1 or 0.2. This criterion is even more needed for the justification and the stability of the computed variance estimates. A computing formula for C_x is given below.
7. Paired selections, with $a_h = 2$ in all strata (though this is not necessary), are convenient for computing variances. They also allow the utmost stratification for a fixed number ($2H$) of primary selections, while still yielding two units for variance computations from each stratum. The variance of the ratio mean $r = y/x = \Sigma y_h / \Sigma x_h$ may be expressed generally and simply as:

$$\text{var}(r) \simeq \frac{1}{x^2}[\text{var}(y) + r^2 \text{var}(x) - 2r \text{cov}(y, x)] \tag{4.7.3}$$

$$= \frac{1}{x^2}(\Sigma_h dy_h^2 + r^2 \Sigma dx_h^2 - 2r \Sigma dy_h dx_h) \tag{4.7.4}$$

$$= \frac{1}{x^2}\Sigma_h dz_h^2. \tag{4.7.5}$$

The first expression (4.7.3) is quite general for the ratio of two random variables. It is written here in sampling terms, and each of the three sampling variances estimates a corresponding population variance (Var), e.g., $E[\text{var}(x)] = \text{Var}(x)$, the variance of the sample sum x. The derivation depends on a Taylor expansion (or delta method or linearization) for the ratio as function of its simple components, the sample sums y and x.

Expression (4.7.3) is an approximation that has been shown to be quite good for large and moderate-sized samples. A convenient and adequate criterion is a small coefficient of variation for x, $C_x < 0.1$ or 0.2, again.

The second expression (4.7.4) takes us toward convenient computing forms when each of $x = \Sigma_h x_h$ and $y = \Sigma_h y_h$ are sums of variables independently selected from the strata.

The third expression (4.7.5) makes for a simplified form using terms $z_{h\alpha} = y_{h\alpha} - rx_{h\alpha}$, the deviations of the $y_{h\alpha}$ values from values "expected" on basis of the $x_{h\alpha}$ values of the same primary selections. These forms are brief, expressive, and easier to check and use for hand computations. Their simplicity becomes even more advantageous for complex functions of r, discussed below.

The computations can be done directly from either (1) the a_h pairs of $y_{h\alpha}$ and $x_{h\alpha}$ from each stratum in (4.7.4), or (2) the a_h forms of $z_{h\alpha} = y_{h\alpha} - rx_{h\alpha}$ precomputed for (4.7.5). For paired selections $a_h = 2$, hence the alternate simple forms on the right in (4.7.6) may be used. Note also that $x_h = \Sigma_\alpha x_{h\alpha}$, $y_h = \Sigma_\alpha y_{h\alpha}$, and $z_h = \Sigma_\alpha z_{h\alpha}$.

$$dx_h^2 = \frac{a_h \Sigma_\alpha x_{h\alpha}^2 - x_h^2}{a_h - 1} = (x_{h1} - x_{h2})^2$$

$$dy_h^2 = \frac{a_h \Sigma_\alpha y_{h\alpha}^2 - y_h^2}{a_h - 1} = (y_{h1} - x_{h2})^2$$

$$dy_h dx_h = \frac{a_h \Sigma y_{h\alpha} x_{h\alpha} - y_h x_h}{a_h - 1} = (y_{h1} - y_{h2})(x_{h1} - x_{h2}) \qquad (4.7.6)$$

$$dz_h^2 = \frac{a_h \Sigma z_{h\alpha}^2 - z_h^2}{a_h - 1} = (z_{h1} - z_{h2})^2. \qquad (4.7.7)$$

Because these forms are complex, mistakes can be made easily and computing $\text{deft}^2 = \text{var}(r)/(s^2/n)$ serves usually as a good check for mistakes. Compute $s^2 = [x\Sigma_j y_j^2 - y^2]/x^2$ when $y = \Sigma_j y_j$ and $x = \Sigma_j x_j$ is a case count. If these variables are weighted, then $s^2 = [(\Sigma_j w_j x_j)\Sigma_j w_j y_j^2 - (\Sigma w_j y_j)^2]/(\Sigma w_j x_i)^2$ and n is an unweighted case count; deft^2 includes the effects of weighting as well as all complexities of clustering and stratification. Often $r = p$ is a proportion, and $s^2 = p(1 - p)$, disregarding the factor $n/(n - 1)$.

For the coefficient of variation of x, to check against both bias in r and instability in var(r), use $c_x = [\sqrt{(\Sigma\, dx_h^2)}]/x < 0.1$ or 0.2.

The many uses of ratio estimators may be discussed under three headings. First, these "combined" ratio means $r = y/x$ may be used with some adjustment factor K to yield more accurate estimators Kr. The factor K may be a "raising" or "expansion" factor to estimate an aggregate, or an adjustment factor for a better mean. These factors may reduce biases of measurement and especially of nonresponse or undercoverage and these may be of great practical utility. They may also yield reductions of sampling variances, and those are more easily displayed in formulas. Incidentally, the word "estimator" is used in statistics to call attention to expected or average performances of statistics like Kr, rather than to the actual result of a specific estimate in a single sample, whose result is subject to variation and to unknown errors. The *adjustment factor K may take several specific forms.*

1. The estimator $Nr = N(y/n)$ *when the population size* N is known from outside sources can be a much more accurate estimator than $Fy = y/f$ based on the sampling fraction f, when y and n are subject to *similar* biases of undercoverage and to sampling errors. Its variance

$$\mathrm{var}(Nr) = \frac{N^2[\mathrm{var}(y) + r^2\,\mathrm{var}(n) - 2r\,\mathrm{cov}(y, n)]}{n^2} \tag{4.7.8}$$

can be much less, with high $\mathrm{cov}(y, n)$, than $\mathrm{var}(Fy) = F^2\,\mathrm{var}(y)$, and the bias for (Nr) may also be much less than for (Fy).

For the same reason r would be a more accurate estimator of the mean $\bar{Y} = Y/N$ than $Fy/N = y/Nf$ would be. This is due to the variable size on which the sample y is based in complex samples. The situation would be entirely different in samples based on fixed sample sizes n_f.

2. These advantages may hold even when $\tilde{N}$ is an estimate, also subject to error, so that for the product $\tilde{N}r = \tilde{N}(y/n)$:

$$\mathrm{var}(\tilde{N}r) = \tilde{N}^2\,\mathrm{var}(r) + r^2\,\mathrm{var}(\tilde{N}) + 2\tilde{N}r\,\mathrm{cov}(\tilde{N}, r). \tag{4.7.9}$$

If the $\tilde{N}$ is based on a large sample, the last two terms may be negligible, compared to $\tilde{N}^2\,\mathrm{var}(r)$. If the bias of $\tilde{N}$ is small, this may be a better estimator than Fy. This (4.7.9) is the variance for a product of two random variables (4.7.17).

3. In the ratio $r = y/x$, the x is sometimes some other positive variable, rather than a case count. Then, if the value X is known from outside sources, we may have the estimator $X(y/x)$ similar to case 1 above. With a stable (large sample) value $\tilde{X}$ we may have $\tilde{X}(y/x)$ similar to case 2 above. Furthermore, we may also consider an adjustment $\tilde{X}$ for the mean (y/x) to produce a better mean. For example (y/x) may be a "calibration", from a quality check (evaluation) with better measurements, on a subsample from the larger sample that produced the mean $\tilde{X}$. Conversely, we may regard y/x as our main product and $\tilde{X}$ the result of a first phase of screening from a larger sample, used for reducing the variance:

$$\mathrm{var}(\bar{X}r) = \bar{X}^2\,\mathrm{var}(r) + r^2\,\mathrm{var}(\bar{X}) + 2\bar{X}r\,\mathrm{cov}(\bar{X}, r). \tag{4.7.10}$$

If the two samples are independent, the last term vanishes. If the $\bar{X}$ is known without error (from a census), only the first term remains.

Second, the "combined" ratio estimator, $r = \Sigma y_h/\Sigma x_h$ is often compared with *several alternative estimators*.

4. One of these alternatives is the "separate" ratio estimator $r_w = \Sigma W_h r_h = \Sigma W_h y_h/x_h$. The "combined" r uses the ratio of two sums,

each based on the entire sample, whereas the "separate" r_w uses the weighted sums of separate ratios r_h within each stratum. The W_h are weights based on "knowledge" beyond the sample data, and the r_h within strata may be unstable if the x_h are small. If the W_h are accurate (unbiased and precise), the *separate ratio estimator* may be used because it has a lower variance,

$$\text{var}(y_W) = \text{var}(\Sigma\, W_h r_h) = \Sigma_h \frac{W_h^2}{r_h^2}[\text{var}(y_h) + r_h^2 \,\text{var}(x_h) - 2r_h^2 \,\text{cov}(y_h x_h)], \tag{4.7.11}$$

than the combined estimator. This lower variance is noted in the literature and it also resembles "poststratification" estimators [Cochran 1977, 6.10, 5A.9]. It is used, I believe, more to correct for biases of nonresponse and undercoverage [USCB 1978, Ch. 5]. Its use in small samples is limited, because if many strata are needed for effective multivariate control, the ratios r_h (or some of them) can be unstable.

5. We may barely compare here *two interesting alternatives* that are described in the literature [Hansen, Hurwitz 1953, 11.2; Kish 1965, 13.2B; Cochran 1977, Ch. 7]. The *regression mean*,

$$\bar{y}_{\text{reg}} = \bar{y} + b(\bar{X} - \bar{x}) = \bar{y} - b\bar{x} + b\bar{X}, \tag{4.7.12}$$

represents a correction of the survey mean $\bar{y}$ with the factor $b(\bar{X} - \bar{x})$ where b is the coefficient of regression of the survey variable Y on the (control) "ancillary" variable X, whose known population mean $\bar{X}$ and sample mean $\bar{x}$ can be used for the correction. This may be viewed as a special case of the more general *difference mean*:

$$\bar{y}_{\text{diff}} = \bar{y} + k(\bar{X} - \bar{x}) = \bar{y} - k\bar{x} + k\bar{X}, \tag{4.7.13}$$

which uses some constant k for the adjustment for a control variable X; whereas the regression mean uses the regression coefficient b. For this comparison the combined ratio mean may also be viewed as a special case of the difference mean with $k = r = \bar{y}/\bar{x}$:

$$\bar{y}_{\text{ratio}} = \bar{y} + (\bar{y}/\bar{x})(\bar{X} - \bar{x}) = \bar{y} - \bar{y} + (\bar{y}/\bar{x})\bar{X}. \tag{4.7.14}$$

Third, the forms used for the ratio means are also useful for *related more complex* statistics (7.1).

6. Differences between ratio means are used constantly for comparing means (percentages) from complex surveys:

$$\begin{aligned}\operatorname{var}(r - r') &= \operatorname{var}(r) + \operatorname{var}(r') - 2\operatorname{cov}(r, r') \\ &= \frac{1}{x^2}\Sigma\, dz_h^2 + \frac{1}{x'^2}\Sigma\, dz'^2_h - \frac{2}{xx'}\Sigma\, dz_h dz'_h. \end{aligned} \tag{4.7.15}$$

The covariances come from using the same sampling units for both sample means; for independent means r and r' the covariance term vanishes. Using the dx and dy forms would result in four variance plus six covariance terms, instead of the three above.

7. The ratio of two ratio means (i.e., "double ratios") may be useful, e.g., lung cancer rates for smokers over nonsmokers (7.2A). Its variance resembles the variance for r (4.7.3):

$$\operatorname{var}\left(\frac{r}{r'}\right) = \frac{1}{r'^2}\left[\operatorname{var}(r) + \left(\frac{r}{r'}\right)^2 \operatorname{var}(r') - 2\left(\frac{r}{r'}\right)\operatorname{cov}(r, r')\right]. \tag{4.7.16}$$

Many more complex indexes, e.g., $\Sigma_g(r_g/r'_g)$, can also be constructed and utilized [Kish 1965, Sec. 12.11; also (6.6C) and (6.6D)].

8. The product of random variables has been used above and the variance for (yx) is

$$\operatorname{var}(yx) = x^2\operatorname{var}(y) + y^2\operatorname{var}(x) + 2yx\operatorname{cov}(y, x). \tag{4.7.17}$$

9. "Relvariances," i.e., relative variances, the squares of the coefficients of variation, allow for very symmetrical comparisons, easy to remember:

$$\begin{aligned}\frac{\operatorname{var}(y/x)}{(y/x)^2} &= \frac{\operatorname{var}(y)}{y^2} + \frac{\operatorname{var}(x)}{x^2} - \frac{2\operatorname{cov}(y, x)}{yx} \\ \frac{\operatorname{var}(yx)}{(yx)^2} &= \frac{\operatorname{var}(y)}{y^2} + \frac{\operatorname{var}(x)}{x^2} + \frac{2\operatorname{cov}(y, x)}{yx} \\ \frac{\operatorname{var}(r/r')}{(r/r')^2} &= \frac{\operatorname{var}(r)}{r^2} + \frac{\operatorname{var}(r')}{r'^2} - \frac{2\operatorname{cov}(r, r')}{rr'}. \end{aligned} \tag{4.7.18}$$

CHAPTER 5

Samples and Censuses

The government are very keen on amassing statistics.... But you must never forget that every one of these figures comes in the first instance from the village watchman, who puts down what he damn pleases. Sir Joseph Stamp.

5.1 CENSUSES AND RESEARCHERS

We have explored the relative advantages of survey sampling compared with those of experiments and of controlled observations (Ch. 1). Later, survey sampling was compared with the complete census of a single local site or a few of them (3.1). Now we shall compare and relate sample surveys to complete censuses of national and of other large populations. These three different comparisons and relationships all have specific practical contents, hence their treatments have been in entirely separate contexts (Fig. 1.3.1).

For this present comparison, consider primarily complete decennial censuses of the national population; but large-scale sample censuses, say 10 percent, would yield similar comparisons and relations. Also, censuses of large states or cities would yield similar comparisons. Furthermore, though we focus on population and housing censuses, we may also consider censuses of farms, business, industry, schools.

Data from registers of the population and from other sources of administrative records are also alternatives to censuses and to samples; and all three can be combined with them for joint uses (5.3F). However, we cannot deal with them adequately here because they are so specific and diverse, and instead refer to other sources [United Nations 1962, NCHS 1980, US Dept Commerce 1980].

I must raise also the question of motivation. In addition to statisticians and technicians in the census bureaus and in national statistical offices, why should researchers outside those offices concern themselves with these comparisons and relations? Let me list several reasons to motivate the interest of social scientists in both the value and the shortcomings of censuses.

1. Researchers often *use the results* of censuses and they should be aware of the limitations, errors, and problems of censuses. Many use only the published results, but some also use census data in more direct forms, especially since the existence of public-use tapes of samples from censuses. They may have to face the choice of using or sometimes even of conducting sample surveys rather than accepting census data that are obsolescent; or seemingly inaccurate; or restricted in depth, richness, and relevance.

2. Researchers may be able to *influence the content*, form, and accuracy of the census schedules and procedures. To perform that role they should appreciate the problems and limitations as well as the advantages of censuses. Their input may seem even more relevant for the sample supplements, which have become common in recent years, than for the core questions of the complete (100 percent) censuses.

3. Researchers may be able to *improve the schedules and procedures* for censuses and for the samples conducted to measure the errors of censuses, to evaluate and to improve them. The methodological skills, experience, and theories of social researchers can be used by the statistical offices if their advice is informed of the practical problems of census taking.

4. Researchers may consider *using or conducting a complete census of a restricted site* as an alternative to a national sample survey (3.1). Several dimensions and criteria are involved in this decision, as well as in the next one.

5. Researchers may consider *taking complete censuses of organizations*, schools, and other institutions. Such large-scale effort may benefit from advice and help from census takers acquainted with the problems and opportunities of large-scale collections of data. The advantages in greater detail and in completeness of a census of a smaller population may be contrasted with the greater breadth and depth possible in a sample survey of a larger population.

6. Censuses have developed a variety of uses for samples to supplement or evaluate or improve censuses (5.3). Researchers should explore *the possible uses of these methods* to improve their own sample surveys also. Some of these methods may be too expensive for realistic use with small samples, but others may be more feasible and justifiable, especially for larger and more expensive surveys.

For this chapter I have made heavy use of two earlier papers [Kish 1979; Kish and Verma 1983]. These contain references to other papers on censuses with deeper content than is possible here and are intended for readers who need to work with methods that are only briefly discussed here.

Finally, I would add here a brief remark about the high costs of decennial censuses, which sometimes get adverse reactions from the press and the public. The 1980 censuses of the United States cost about one billion dollars. But that comes to only $4 per capita for the population of 250 million, which amounts to about one-half hour of work at median wages decennially. Interestingly, that half hour at median wages seems to hold fairly well, I believe, for very diverse decennial censuses of other nations also. We should add to those open costs the hidden cost of the time spent by a population of respondents. But then we should also subtract from those costs the benefits of the nation's population partaking in the "ceremony" of assessing its individual and its collective statuses.

5.2 SAMPLES COMPARED TO CENSUSES

In comparing sample surveys to complete censuses we shall note their *relative* advantages and weaknesses: each seems stronger where the other is weaker. Thus we may view them in competition with each other, but also emphasize their complementary natures. Then we note how they can be used to aid each other and also used jointly in postcensal estimates for small areas (5.3F).

These comparisons concern covering the same large national population either with a complete census or with a much smaller, hence also less expensive sample. Different dimensions are involved in comparing for similar costs a census of one restricted community (or a few) with a national sample (3.1) and in comparing a longitudinal study restricted in space with a one-time cross section of a national population (6.1). The comparisons also have different dimensions for restricted populations, e.g., the national population of a professional society or the total population of a small city-state or small island-nation. The contrast between samples and complete censuses is more striking and decisive generally for large, national populations, and that is our central concern here.

The United States has almost 100 million households, and a sampling rate of 1/1000 would yield a sample of 100,000, and even a rate of 1/100,000 yields $n = 1000$. Censuses are entirely different in the scale of operations involved, and in the consequent differences in methodology and the practical conditions of data collection. Complete enumeration of the entire population requires mobilization of financial and human resources on a large scale, which cannot be sustained for a prolonged period or repeated frequently. The need to deploy a large—hence a less well trained and less closely supervised—field force means that the type of information which is appropriately collected in a census, while extensive in coverage, has to be relatively simple in content. Simplicity is also necessary to keep the volume of data to be processed manageable. Complete censuses are nevertheless relatively expensive and slow, and even with today's modern, efficient procedures it can take four years to get most of the census data into the hands of users. These are the basic reasons for not taking censuses more often, or with greater depth and richness of data.

Therefore the primary objective of a census is typically to obtain a detailed and complete picture of the number (size) and basic structural and related characteristics of the population, and to provide as much detail as possible for small domains and especially for local areas. For example, the population census provides information on the size, age–sex composition, geographic distribution, and most basic demographic and socioeconomic characteristics of the population.

By contrast, inquiries confined to samples of the population can, by virtue of their smaller sizes, be designed to obtain a wide variety of complete data for studies of interrelationships and of changes. Such data are not gathered in complete censuses: attempts to do so would result in very high costs and, even more important, in low quality. (This is illustrated by misguided attempts in some countries to collect data on disability or on abortions in population censuses.) Furthermore, sample surveys can be tailored flexibly to fit a variety of needs with appropriate methods of collection. Choice of timing, of respondents, and of methods can be suited to the needs of data collection. The content of the study population can be better controlled and directed toward the specific survey aims; such flexibility may be prohibited by the public aspects of the census. Sample surveys are much cheaper, and they can be made much more timely. They can be repeated more often to provide information on rapidly changing or fluctuating variables.

The major limitation of sample surveys is their inability to provide sufficient detail for small domains and especially for local areas. This is the principal reason for the continuing utility of complete censuses. Though even here there are certain important qualifications [Hansen et al. 1961, Waksberg 1968, Kish 1979]. Furthermore, censuses can often (though not always or necessarily) obtain better coverage and response rates than sample surveys. This is partly because it is less difficult to check complete coverage than sample coverage, but mostly because of the credibility aroused by the public relations campaigns for censuses. Thus censuses can obtain better representation with greater coverage, because of better frames, than samples.

Samples usually depend on information from external sources not only for frames for sample selection, but also for more precise ratio estimation and

TABLE 5.2.1. Eight Criteria for Comparing Three Sources of Data[a]

Criteria	Samples	Census	Administrative Registers
Rich, complex, diverse, flexible	***		
Accurate, relevant, pertinent	*		?
Inexpensive	*		***
Timely, opportune, seasonal	**		*
Precise (large and complete)		*	*
Detailed for small domains		**	*
Inclusive (coverage), credible, P.R.		*	?
Population content	**	*	

[a]From Kish 1979.

similar methods (4.7). Estimates of population totals obtained directly from the sample, by inflating the sample totals, would often suffer severe underestimates (downward biases) due to noncoverages by the sample; thus they are seldom used. Those totals can be estimated more precisely as ratio estimates of sample means multiplied by totals from a census or from other sources. However, samples of moderate or even small size can often yield with adequate and useful precision the estimates for means, proportions (percentages), ratios, etc. Those estimates and comparisons of means, and more complex analytical statistics, are often the main purposes of surveys and some can also serve as bases for estimating totals.

The eight criteria in Table 5.2.1 represent my compromise between brevity and completeness to cover adequately the needs of diverse situations. In any situation the researcher must judge how well each source meets those criteria. This list may help to avoid a choice by only a single criterion. Asterisks (*) are used to indicate the relative advantages of each source, according to each criterion. The (?) for administrative registers and records refers to their extreme differences for diverse variables in different situations. Birth and death records, utility records (telephone, electricity), tax records, etc. can be accurate or bad; and this remark about accuracy goes for both relevance and coverage. Their use can, of course, be very inexpensive (***)—when they are available—because their costs were borne by other users. But they seldom have the rich, diverse data needed for social research; and the population content may be limited to households, to their "heads," or to undefined portions of the population. The striking feature of the table is the complementary nature of samples, which dominate five criteria, and of censuses, which dominate the other three.

5.3 SAMPLES ATTACHED TO CENSUSES

5.3A List of Sampling Applications

The literature of survey sampling is oriented to designs for distinct and separate surveys. But sampling methods can also be applied to samples connected with censuses, and these samples have special features because of their double roots. They share methods, techniques, and theory with survey samples. But their connection with censuses gives them both special functions and special advantages in funds and in resources. Hence they often have large sizes; especially in class 1 of Table 5.3.1 (5.3B). They also share with censuses some inflexibilities in timing, in the contents of their populations, and especially in the restrictions on data that official censuses can afford to collect without jeopardizing the wide cooperation they need and get.

The list of available methods should be inspected for their possible utilization and practical utility. Any of the methods may be placed on one of four levels of availability for any specific situation, referring to either actual or potential use: (1) successfully used already; (2) used but not successfully or adequately; (3) not used though available; and (4) resources not now available. Needed methods at levels 2, 3, or 4 should be examined for possible transfer to level 1. But methods that are not needed badly enough currently belong elsewhere. On the other hand, methods that would be needed, but are not now available, should deserve special considerations from appropriate technicians.

Most of the methods listed here have appeared in the literature on samples connected with censuses. Those methods are described in some detail, with justifications and procedures, in many references to which there are several guides [UN Statistical Office 1980, Gurney and Manno 1971, Kish and Verma 1983, Kish 1979]. Most references have detailed and technical treatments of only one kind of application. Our treatment here is much briefer and superficial, but it covers 16 kinds of applications in five classes of purposes (Table 5.3.1).

There has been great recent growth in the use of samples connected with censuses. In the 1980–81 round of censuses most countries have used one or more of these applications; and probably a majority have used classes 1 and 2 [UN 1980]. Their uses are probably more uneven than would be dictated by genuine differences in objective situations: there seems to be a great deal of arbitrariness about which methods have been used and where, depending on the choices of individual statisticians and on the decisions of ministers. This presentation aims to help future choices and decisions. I hope especially that

TABLE 5.3.1. Samples Connected with Censuses

1. Sample enumerations to supplement complete censuses:
   (a) Obtain richer, more diverse, detailed, deeper data
   (b) Reduce costs of collection and of tabulation
   (c) Obtain more accurate data, perhaps with special enumerators
   (d) Reduce aggregate social burden on respondents
2. Samples added to complete censuses to evaluate and to improve them:
   (a) Evaluation studies of content (Postevaluation studies)
   (b) Coverage checks; dual coverage
   (c) Pilot studies of questions and techniques before the census
   (d) Quality control of individual enumerators, coders, processors
3. Samples from census records, microfilms, tapes:
   (a) Early (advanced, preliminary) tabulation and releases
   (b) Complex, multivariate analyses of relations
   (c) Public-use tapes for further, deeper analyses (without identification of respondents)
4. Census as auxiliary data for samples:
   (a) Data for selections: measures of size, stratifiers, maps of enumeration areas; seldom addresses or names
   (b) Data for improved estimation with ratios, regressions
   (c) Samples added to censuses to serve as bases for continuing surveys
5. Joint uses of several sources:
   (a) Current estimates for local areas and small domains
   (b) Rolling (rotating) monthly samples of 1/120 (weekly 1/520)

it can stimulate the combined use of several methods, where feasible. The list is meant to be complete, and the descriptions are brief but, I hope, adequate as reminders for readers who are or would become acquainted with the methods and can look up details elsewhere. My five classes for 15 methods are somewhat arbitrary but useful, I hope. The classification is by different purposes, and notes will be given later about the different methods that may be used for them. The samples may be timed to occur before, during, or after the census. The sample schedules may be added to the census schedule or be separate from it; and they may be independent from or dependent on the census schedule. The sampling units may be households, enumeration areas (EAs), or administrative units. Some of the possible combinations of all those possible procedures and purposes are better than others.

Administrators and the technicians of statistical offices should have joint chief responsibility for the contents and methods of censuses, under the general guidance of public bodies and officials. But social researchers and other users outside that definition need not be mere passive receivers of data.

For using census data, for analyzing and reworking them, they need to be aware of the problems and limitations and of the possibilities inherent in those data. Furthermore, some researchers can make useful contributions to improving the quality and enlarging the scope of census results.

5.3B Samples as Supplements to Censuses

On complete censuses each question is expensive, because it is multiplied by the sizes of the populations (of households, persons, farms, etc.). Hence complete censuses should be brief and simple, sometimes with precoded or easily coded items, (and sometimes now with self-enumeration forms) to save costs of collection and tabulation and to reduce the respondent burden on society. However, ever more diverse data are being obtained with samples that are portions of the entire census. These samples are substitutes for complete censuses; hence they tend to be large, ranging from perhaps 1:100 to 1:4 of the complete census. These result in very large samples compared with the usual survey samples, yet they can achieve most of the savings that sampling from censuses can yield. It may even be desirable to design a large sample (1:4) for items needed in great detail, plus a much smaller sample (1:100) for more difficult items, which can also tolerate less detailed results. Sampling reduces the total cost for these data, yet it may also yield higher accuracy, especially if special enumerators can be trained for them. On the other hand, compared with ordinary surveys, large sizes are facilitated by the availability of funding and the efficiency of census operations, which are cheaper per schedule than sample surveys.

Concerning methods of sampling and data collection, several choices must be made, and they are interrelated. (1) A selection rate may be applied to sample households from all enumeration areas (EAs), or the sample may consist of entire EAs. (2) The sample may use the regular enumerators of the complete censuses, self-enumeration by respondents, or special enumerators trained for the sample. (3) The sample schedule may be jointly completed with the core census schedule, done as a separate operation, or perhaps done jointly with another operation, such as 2a (evaluation) or 4c (continuing survey).

For surveys attached to the census, the sample may be spread to all EAs; a sample of EAs may be selected for complete enumeration within EAs; or it may involve a two-stage selection of EAs, then of households. The choice of the design is influenced by a number of interrelated factors: the size of the operation and the required degree of detailed breakdown of the results; nature (complexity) of the supplementary information to be collected; size and nature of EAs; travel conditions; type of enumerators available for the census and required for the attached survey; how often the households can be visited and the related considerations of time, cost, and respondent burden;

whether the attached survey replaces or is additional to the ordinary census operations in the sample areas. When several samples are used to get different data, there is a conflict concerning spreading them over many different units versus concentrating them all in the same units. Spreading the schedules avoids the concentration of respondent burden, but concentrating them reduces costs and yields more information on relations between sets of variables.

For simpler items that can be combined with the basic census enumeration—during a single visit, using the ordinary census enumerators—the sample can be easily spread over all census areas. The more complex and specialized the inquiry, and especially if specialized enumerators are involved, the more advisable it becomes to concentrate the inquiry to a sample of EAs. This can be more advisable also when the objective is to produce results at national or major domain levels, rather than at the level of small domains or local areas. Selection of complete (compact) EAs has the advantage of simplicity and lower costs; it is particularly appropriate when specialized procedures and enumerators are used, or when the survey replaces the ordinary census operations in the sample areas. However, this concentration also increases variances of the estimates, which would be most serious for small geographic domains, though it may not be critical for estimates for major domains and for crossclasses well distributed over different areas.

5.3C Samples to Improve Censuses

Whereas the preceding supplements substitute for and resemble censuses, samples added to improve censuses differ from them and are usually smaller, perhaps from 1:100 to 1:1000, or even to 1:10,000. Postenumeration studies (PES) have been used to evaluate and check the quality of census enumerations, to estimate their biases, and to measure response errors. In some versions the PES enumerators are given the census responses for their sample cases; then they use them to get the "best" answers with more and better questions. In other versions, the PES interviewers are kept ignorant of census responses in efforts to get PES responses independent of them. However, independence is not complete, because the respondents have not entirely forgotten the census interview. Reconciliation of the pairs of responses for a "best" answer can come later.

Checks for completeness of coverage would usually utilize the first version: the check enumerators would have a list of units in defined areas, and then try to find missed units. Checks for coverage from data independent of the census are possible, but less likely. On the other hand, sample studies using the techniques of "dual coverage" [Marks, Seltzer, and Krotki 1974] for estimating undercoverage are possible where lists of households (or other units) are available from some source (such as registers). The procedures for (and the noncoverage from) this other list should be quite different from the census methods for the technique to be fairly effective. The undercoverage is measured from the differences of units covered by one list but not the other.

Instead of a PES done after the census, a sample of high-quality enumeration may be done simultaneously with the census. A sample of EAs may be covered with better methods, better enumerators, longer questionnaires, instead of the census methods used in the remainder of the country. The extra expense is less than with double coverage of sample areas, and the respondent burden of double interviews is avoided also. The contrast of these check areas with areas covered by census methods yields estimates of the net bias. These estimates of the net differences from the sample/census comparisons are free from the bias of memory to census responses, but they have higher sampling errors than with double coverage. Also, the method lacks estimates for the kind of errors that may be obtained from double coverage of the same households and individuals. But on the whole this is a simpler and cheaper method. The sample areas should be selected with careful matching (stratification) of control areas. The sampling units for quality checks are more likely to be EAs or administrative districts than households, because these would be inconvenient and difficult to administer.

Evaluation surveys are designed to check the average quality of the census and of its major components. Quality control and correction of individual enumerators are different matters; they need specific treatment suited to actual field conditions and to procedures of supervision. The quality control of editors and coders in the office is another specialized matter we will not treat here, and they are easier than checks for fieldwork.

5.3D Samples from Census Schedules

Whereas in classes 1 and 2 we discussed sampling of the data collected in the field, in class 3 we are concerned with sampling from the already collected census data. There are three distinct purposes for such samples, and their timing differs greatly; hence they need different methods of selection.

Where early tabulations and releases are wanted, it is convenient to base them on selections of entire EAs (or even administrative districts) in accord with the system of returns from the field collection. The selections should be predesignated and speeded along. They should represent good and valid samples, not merely the first arrivals, which are bound to be biased portions of the population.

Continuing advances in both statistical and computing methods have made it both desirable and possible to conduct more complex analyses of census data, and demands increase for deeper multivariate analyses of relations. For some of these it is convenient to select samples from the entire census to reduce computations, though this need for sampling may be reduced with faster machines and better programs. The analyses can vary in nature, scope, and timing. They are usually done from tapes in the statistical offices to preserve the confidentiality of the data.

Public-use tapes are also prepared from census tapes for the use of researchers. Data that could identify individuals are removed from the tapes, and random selections help greatly to prevent identification. Samples of households are preferred for these uses; spreading the sample reduces the level of sampling errors, and it also facilitates the estimation of those errors by avoiding clustering. Households are easier to select than persons, and they provide samples of persons, families, and households. The clustering of individuals in households matters little in analyses, which seldom group multiple members of the same households into the same cells. Such public-use tapes are gaining in use and several countries are preparing them. The spreading availability of computers and related skills is chiefly responsible for this growth. Furthermore, public-use tapes are also being prepared from schedules of old censuses for historical analyses. It is also true (and sad) that the releases of "current" census data may need several years, making their analysis somewhat "historical" for rapidly changing variables.

5.3E Censuses as Auxiliary Data for Samples

When comparing the advantages of sampling, we must remember that good samples benefit greatly from being based on and aided by census data. Those aids and bases are especially important for national and other large populations and especially for samples of household members. Censuses provide the chief sources for measures of size of sampling units. They are also the chief sources for stratifying variables. Seldom or never can we find other sources with the detailed and complete population coverage that the census provides at low relative cost. Furthermore, census data, maps, and boundaries for EAs serve in many samples as important sampling units. The EAs usually serve as either primary or intermediary sampling units, and the sampler must provide the final list of dwellings (households). Addresses and names from censuses can seldom be used, because of both their confidentiality and their obsolescence.

Census data can also be used to improve statistics, especially through ratio and regression estimates (4.7). When used properly, their inaccuracies, differences, and obsolescence need not cause the biases associated with naive and improper uses.

Some of the samples described earlier could be used directly as bases for continuing surveys. This is especially true for samples of EAs used for supplements (class 1) or evaluations (2a). These samples could then have direct links with the census. This is merely a proposal for which I know of no application at the present.

5.3F Joint Uses of Samples, Censuses, and Registers: Estimation for Small Domains

Census data are usually obsolete, data from registers inadequate, and sample data lacking in detail, especially for local areas. Since the strengths and weaknesses of the three sources are complementary, it seems reasonable to try to combine the strengths of the three sources to obtain estimates for small domains, especially for local areas; estimates that are current, pertinent, and accurate. To the general needs of researchers have been added the needs of social planners, of administrators, and of policy makers for valid, current data for small domains and local areas. *Local area estimation* has become a fast-developing field, being pushed by increasing demands, and simultaneously pulled along by new developments in computing technology and new statistical techniques. These problems of "*postcensal estimates*" are treated currently as technical problems for estimates of the total population in small local administrative areas, with a new, large, but specialized list of publications; a few references can be the key to the longer list [Purcell and Kish 1979, 1980; Heeringa 1981; Rao 1986].

Estimates for small domains serves as a third name for this new field and is more indicative of its future interest for social researchers.

We may also consider future designs to obtain the detailed data of censuses from rotating samples. For example, a rotating monthly sample of 1:120 can cover the nation in 10 years. If it is necessary to measure monthly changes, we can have samples of 1:60 with 50 percent overlaps. The collection period may be spread over the entire month or be confined into representative weeks [Kish 1981].

The joint use of registers with sample surveys is becoming so successful that Denmark substituted those estimates for local areas and omitted its 1980 census. Other Scandinavian countries, with excellent population registers, are considering following that example to save the cost of the 1990 censuses. But most countries will probably need censuses for many years yet.

CHAPTER 6

Sample Designs Over Time

Consequently the results obtained at a single place and in a single year, however accurate in themselves, are of limited utility either for the immediate practical end of determining the most profitable variety, level of manuring, etc., or for the more fundamental task of elucidating the underlying scientific laws. F Yates and W G Cochran, The Analysis of Groups of Experiments.

. . . the race is not to the swift, nor the battle to the strong, neither yet bread to the wise, nor yet riches to men of understanding, nor yet favour to men of skill; but time and chance happeneth to them all. Ecclesiastes, 9.11.

There have been many articles published, as well as chapters and books, about separate problems and individual aspects of the time dimension in social research and related fields. Some of these have appeared within the context of survey sampling, some in the literature of experimental designs, and many concerning observational studies. Those treatments, whether theoretical or empirical, usually explore some single aspect or a few limited aspects of the time dimension, and even within a single aspect only two or three alternative categories are compared in detail and depth. For example, discussions in the sampling literature tend to concentrate on designs for net changes versus the usual static cross sections, and even here research purposes are subordinated to the design aspect. Other examples: Problems of longitudinal studies, of nomads, of retrospective data, etc., are all discussed as separate problems, in an *ad hoc* manner.

On the contrary, here we try to be comprehensive: to explore jointly all the diverse aspects peculiar to the time dimension, and to uncover the connections between all those aspects. Also, under each aspect we attempt to list all the alternative available categories. These bold claims of comprehensiveness should awaken the attention of the readers and their instincts for hunting counterexamples. Is this list of aspects and of alternatives within aspects complete and useful; are these lists both sufficient and necessary?

But that immodest claim can be softened with modest and necessary admissions. First, most of those aspects and their alternative categories may be familiar to most readers and few or none may appear entirely new; only bringing and linking them together is novel. Second, the boundaries of some categories may seem arbitrary. Third, the values of the categories depend on the reader's judgment of their utility in diverse situations. Fourth, our exploration must be brief here: Our list can serve only to remind readers of all aspects and of the available alternatives under each; to bring to mind what they already know of each; and to encourage them to consult references under each. (Note that Section 3.6 deals with times curves of response and Section 3.5 with biases due to time T1 to T4.)

This overview of all temporal aspects may help the researcher find a better combination of the many alternatives for an integrated design. Several good references may best serve as background and as comprehensive guides to the analysis of longitudinal studies [Janson 1981; Schulsinger, Mednick, and Knop 1981; Duncan and Kalton 1986; Goldstein 1979; Wall and Williams 1970; Harris 1963]. These and my many references cannot keep up with the veritable explosion of new publications on longitudinal studies. The reasons for that vast recent expansion come from the recent growth of empirical work, of many new longitudinal and repeated surveys in social research, almost absent in earlier years. That growth is being accompanied by the rise of new methodological and computing tools. I offer three explanations for not covering more fully the new publications. First, many of them deal with analytical tools, more than with designs that are covered here. Second, most of the others deal with the details of measurement and administration, also beyond my coverage. Third, even if I could cover them in 1986, the rapid new developments would soon make them obsolete. In any case, many of the following ideas about design were fresh and are useful now.

6.1 TERMINOLOGY AND CONCEPTS

6.1A Terminology

My aim is practical rather than pedantic: to make the diverse terms commonly used in this field serve us better by having clearer distinctions between them; to strive for one-to-one correspondence between terms and uses, instead of multiple uses for the same term and multiple terms for the same use. I was guided partly by current usage (which is not uniform), partly by the connotations of words, and partly by the diverse needs of the field.

Repeated surveys denote "similar" observations on the "same" population, but without specifying designs for the overlapping coverage of the

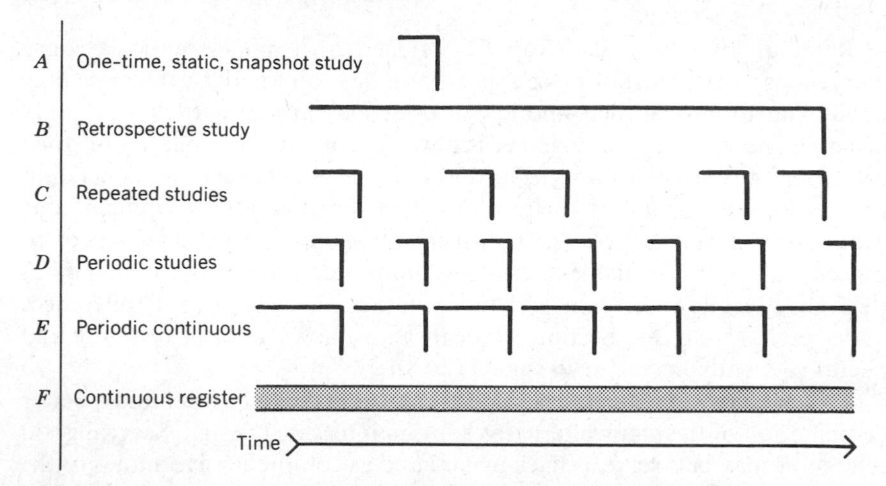

Figure 6.1.1. Designs for covering time spans over populations.

Vertical stems denote collection dates and horizontal bars the reference periods being covered. Retrospective studies cover long periods, perhaps entire lifetimes, from one collection date. Repeated studies may be collected at irregular intervals, may have different reference periods, perhaps ending before the collection date. Periodic studies come with evenly spaced collection periods and with similar reference periods generally. Continuous periodic studies cover the entire time span with contiguous reference periods. Continuous registers cover the entire time span. All of these, except for the one-time "snap shot" A, have been loosely called "longitudinal" studies. However "strictly longitudinal" has been used to denote panel designs for the same elements, as in 5 of Figure 6.1.2. Otherwise the coverage refers to the same defined population, although individuals may change over the span through births and deaths and migrations. For example, repeated studies with C, D, or E can be applied to single communities or to countries, although their inhabitants change continually.

same set of units. *Periodic* surveys refer to surveys repeated at specified regular *periods* over a longer *interval* of time. The "same" population needs identification because populations change over time both in extent and in content (e.g., cities and countries change boundaries); for complex units (families, organizations) changes can be frequent as units (persons, adults) are born, die, and migrate. "Similar" observations must also be defined, operationalized, and collected (Figure 6.1.1).

Overlapping designs refer to covering the same sampling units in repeated periods. The overlapping units may be defined as the elements of analysis (individuals, persons), or they may be larger units, such as area segments. Units such as families, households, composed of distinct elements, present problems of frequent and complex changes. Designs may require either *complete or partial overlapping*; the latter permits gradual changes of the sampling units. In *nonoverlapping* designs the units are changed deliberately for each period (Figure 6.1.2, Section 6.2).

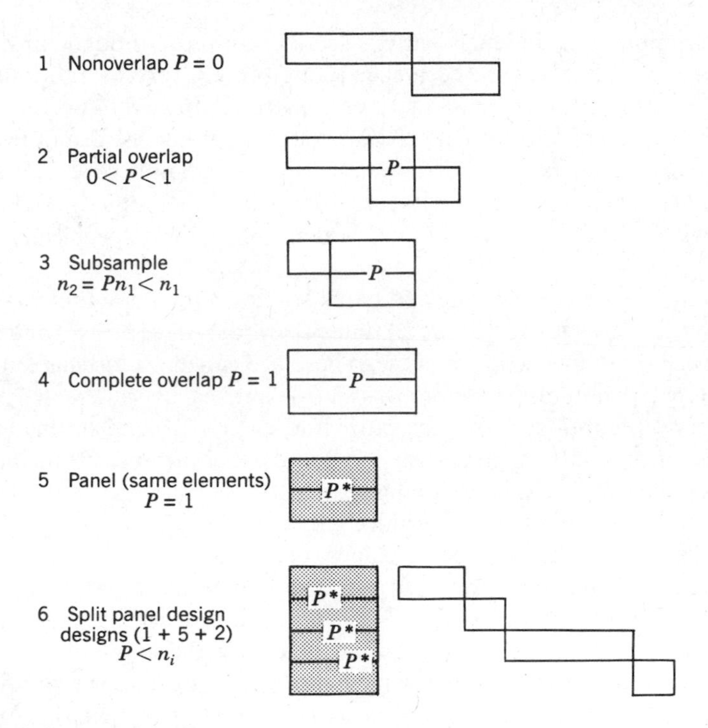

Figure 6.1.2 Designs of overlaps.

The overlap is $P = 0$ for design 1 and $P = 1$ for designs 4 and 5 in two periods. These may refer to clusters of elements, such as area segments, blocks, or towns. In designs 1, 2, 3 period 1 is shown to have larger samples than period 2. Panel for design 5 denotes a complete overlap P^* for the same elements (elementary units, e.g., persons). The designs, shown for two periods, can also be applied to more periods. Design 6 combines a repeated panel with nonoverlaps of different sizes in a design I call split-panel design (6.5); the panel P^* provides partial overlaps $0 < P < 1$ for *any* two periods.

Panel surveys refer to overlapping studies with repeated observations on the same elements, on the same persons. Panels face problems of learning, fatigue, and losses from mortality and mobility; of moving and high locating costs; and of identification for complex units, like families; but they are needed for detecting the dynamics of *gross* (micro) *changes* of individuals (though these get confounded with errors of measurement). On the other hand, for measuring *net* (macro) *changes* of averages it may be easier and clearer to overlap simpler units of sampling (such as area segments) and still retain much of the gains in the variances from correlations. (Some studies have done both: retain segments for clear net changes and follow moving individuals for gross changes.) The gains from correlations are also retained proportionately in partial overlaps. Net changes may be measured also with nonoverlapping or independently selected samples, though with higher variances. Panels have also been called longitudinal surveys and follow-ups. But the term *follow-up* has also been used confusingly for overlapping surveys, for samples for quality checks, and for call-backs for nonresponses. And *longitudinal* can also refer to several kinds of repeated surveys; *longitudinal studies* is commonly used for various methods of analysis over time, with *strictly longitudinal studies* denoting panels specifically [Janson 1981].

A third use for overlapping and panel studies is for obtaining *incidence* of new events between two (or more) dates (periods), in contrast to measuring *prevalence* of all events at one time. These are called *multiround* surveys by some (e.g., demographers); or *prospective* studies by others (e.g., health scientists), in contrast to *retrospective* studies that depend on memories or records for past data. Such designs should be panels for measuring individual changes, but can be nonoverlapping studies for net changes. The collection of data on new events is sometimes aided with records (diaries, budgets) kept by respondents, by others, or by machines, etc.

To avoid confusion we need to distinguish three kinds of periods concerning any survey: a *collection period* during which data are collected; *reference periods* defining the data, which may differ greatly for diverse statistics; and *reporting periods*, which can consist of one or more reference periods. For example, the U.S. Census is collected for weeks in April, the reference and reporting periods are April 1 for current data but the preceding calendar year for economic data, etc. In multiround and cumulated surveys the reporting periods are pooled from reference periods (6.6). Reference periods may be as short as a single day or even a minute (in time studies), a week (for employment) or month, or as long as a year (for income).

6.1B Representing Time Spans

Changes in populations and in associated variables occur within the perceived smooth, unidirectional flow of time. We can now briefly compare alternative strategies of temporal designs for social research and sample surveys; usually one of these alternatives is accepted merely by habit or tradition. Probability selection of and averaging over the time dimension has been neglected compared with spatial and other aspects of populations. That neglect may be explained by familiar collection methods: decennial censuses on traditional days; cross-section samples on arbitrary or customary days. Bases for inferences over long time intervals have not been investigated thoroughly.

1. *Unique or special periods* may be accepted from natural forces (e.g., seasons for harvesting crops, for lambs' births, monsoons, etc.). Dates fixed

by laws, rules, and customs—such as Christmas, New Year, fiscal year, month's end, Sundays (or Fridays or Saturdays)—may seem arbitrary, provincial, and temporary, but they are fixed for the population, hence beyond the designs of researchers.

2. "*Typical*" (*representative*) *periods* are commonly used; all too commonly, I believe, in confusion with either uniquely fixed times (1) or proper sampling of time (4). A good example is April 1 for the reference dates of decennial censuses of the United States, because that date was thought to be more "typical" and convenient than many other days and is now traditional. Less convincing seems the choice of the third week to represent each month in the Current Population Surveys [USCB 1978]. However, there are many worse examples of choosing "typical" (representative) periods by judgment in preference to sampling the time dimension; tradition, convenience, and costs may explain their prevalence. Remember that "typical" areas were also commonly used for spatial representation until the recent spread of probability area sampling.

3. *Complete and separate coverage* of all reference periods over the reference interval is a temporal analogue of a complete census over all administrative areas. These yield data for all periods for averages over them and for changes between them—e.g., the yearly survey over all 52 weeks of the Health Interview Survey (NCHS 1958]; different examples arise from time series of some financial data. We can distinguish continuous from discontinuous periods over the entire intervals; the Current Population Surveys [USCB 1978] cover all 12 months over the year, but only a "typical" week to represent each month (E versus D in Fig. 6.1.1). Continuous collection of data is seldom feasible; but reference periods can be, as in multiround surveys, and from these the aggregates and means for entire intervals can be computed. However these naturally raise the possibility of sampling instead of complete coverage of all reference periods.

4. *Sampling over a time interval* can be an alternative to either confining the sample to one or a few "typical" (representative) periods (2) or completely covering the entire interval (3). Models of temporal variation can be made similar to spatial variation: as a target population varies in space so we can consider time as another dimension of variation. Populations vary from year to year and week to week, as they vary among regions and among counties. Probability sampling spread over the population area serves as the accepted strategy to cover and counter spatial variation. But temporal variation can be even more important, especially for cyclical variations, e.g., seasonal, weekly, or even diurnal. Vast temporal fluctuations also occur in epidemics, economic situations, and social and political attitudes, and rapid and widespread changes have become common. To cover and counter these changes either complete coverage or sampling is needed. However, for many

characteristics, which have temporal stability but much spatial variation, spatial coverage may be more crucial.

For covering and sampling entire time spans we must overcome not only the inertia of habit, and tradition, but also some practical obstacles, discussed in (6) below. Philosophical differences also exist, because we accept the spatial boundaries of nations, and of administrative areas, whereas time appears continuous, smooth, and unidirectional. But we accept natural periods of day and year, also cultural spans such as weeks, months, and decades (historians use centuries, dynasties, and ages in hindsight).

In any case, definitions and models of the target population should have temporal as well as spatial dimensions (2.1). When practical constraints, as for decennial censuses, demand a single period, we agree that judgmental selection of a representative period seems better than a random choice. However, when a larger number of periods seem feasible, consider a probability selection of periods (as you would of areas), e.g., a stratified selection of 24 out of the year's 365 days.

Methods for sampling time can be flexible and diverse, similar to samples of spatial dimensions. Selection with probabilities proportional (PPS) to measures of size (MOS) may be used. For example, a sample of clients may be selected from a selection with MOS of institutions of diverse sizes. Thus probabilities are maintained for clients in the population, since large institutions receive large selection probabilities (MOS) but proportionately shorter periods (If all clients entering within short periods must be accepted, the lengths of the periods may be varied inversely to the MOS for institutions.)

5. *Temporal × spatial matrix* for averages (marginals) for both dimensions can be designed with periodic samples. A good example again is the Health Interview Surveys [NCHS 1958] that yield weekly national averages, yearly statistics for small domains, and monthly and quarterly data for larger domains. For an early example see [Kish et al. 1961]. The samples are too small to yield both spatial and temporal details simultaneously, but each period can be designed to sample the entire population area; furthermore, the periodic samples can be so controlled that they cumulate to subtotals (regions) and totals (national) that are balanced (stratified).

In the time × space matrix we cannot select each cell, but we can sample each row and each column with a balanced design. Each weekly row may yield national averages; each spatial column perhaps yearly averages, and monthly averages for larger domains; and each average can be balanced against the other dimension. The design resembles stratified or controlled or Latin-square designs for two variables. Furthermore, in addition to spatial variation, other dimensions—social, economic etc.—may be covered. (See also Section 6.6D for cumulated and rolling samples.)

Sampling the periods may thus be an alternative to complete coverage (3) to achieve greater spatial coverage and to lower cost. The commonly assumed contradiction of "longitudinal" versus "cross-section" designs is false conceptually, but it does refer to cost problems of covering long spans of both time and space; hence typically cross sections tend to sacrifice temporal coverage whereas longitudinal studies are confined spatially. Matrix designs for sampling both space and time with balanced periodic samples may provide a better solution.

6. *Repeated (periodic) surveys versus single periods.* Periodic surveys designed for specified periods can be averaged over their entire interval; similarly, a large survey for one period can be divided instead into several periods and averaged over a longer interval. The advantages of repeated samples need stressing. (a) With repeated samples we may be able to improve quality and reduce costs. Hiring and training a large force of enumerators and clerks for one large survey (or census) can be an expensive and risky input for the short survey effort. (b) Repeated samples can yield statistics for time series to estimate changes and trends (seasonal, secular, etc.) and to detect irregular and sudden variations. (c) Averages and sums of repeated samples can lead to better statistical inference than a single, concentrated, "one-shot" sample: Probability selection of time segments from the entire interval permits statistical inference from the sample to the interval. On the contrary, inference from a judgmental choice of a "typical" period depends on subjective assumptions and is exposed to the vagaries of unknown trends and irregular variations.

On the other hand, the familiar advantages of "one-shot" samples are widely accepted. (d) A specific date may fit some legal or traditional requirement. (e) A complete census or heavy coverage during one short period may be simpler for investigating relations among many variables and units. (f) It may be easier and cheaper to operate, especially for small and widespread samples, than repeated samples spread over a long interval. (g) It permits analysis, presentation and use of results more rapidly after conception and funding [Kish 1965a, 12.5D].

6.1C Collection of Data: Alternatives Over Time

1. *Retrospective data* generally refer to methods based on the memory and report of respondents to obtain data about events, behaviors, and attitudes. Almost all social data are retrospective in some sense, since researchers seldom witness the emergence of data; hence we must use relative frames of reference. Near one extreme are birth histories; mothers can recall and report rather well total number of live births, as well as the number of their living children; less well the total pregnancies, including miscarriages and abor-

tions; even the dates of all births tend to be well reported in some cultures but not in others. We see here the requirements for good retrospective data for longer periods: Events should be rare and important for long recall. The respondent has to be aware of, able to recall, and willing to report the events, and for some data also their timing. For example, respondents may be unaware of or may have forgotten some miscarriages and their timing, and may be unwilling to report abortions or deaths.

For consumer items, dates and data about house purchases may be remembered for years; for one year for autos; but hardly a day for bread or cigarettes. The quality of retrospective data varies greatly for different items, different periods, different respondents and cultures, and different methods of measurement. We cannot hope here even to refer adequately to the vast, confusing literature [Sudman and Bradburn 1973, 1974; Moss and Goldstein 1979; Schuman and Kalton 1982; Gray 1955; Zarkovich 1963]. Retrospective data are and must be widely used because they can yield so much more data for so much less effort than the following alternatives. We must distinguish between respondents (observational units) and the individuals (population elements) in a study; e.g., mothers would be respondents on diseases and vaccinations of their children.

2. *Registers, records, or direct observation* can at times be alternatives to retrospective recall. The preceding problems above of awareness, recall, and response can be translated here into completeness, availability, and errors. Registers refer to data kept for some administrative use (the population registers of Scandinavian countries are the best example); but there are also records of schools and of utilities, income tax returns, birth and death registers, and many others of diverse value and coverage. The data are cheap and continuous, but their extent, accuracy and availability are often greatly limited [Janson 1984; UN 1962; Marks, Seltzer, and Krotki 1974; U.S. Dept. Com. 1980].

On the contrary, records kept by respondents specifically for research objectives may require great expenditure by the project and efforts by respondents, and they may meet reluctance and refusals; but they have been used for studies of buying, income, and other behaviors. Methods for sampling observed behavior with reduced costs and respondent burdens are difficult but have been devised, e.g., machines for sampling home TV operation. Hourly activities have been sampled with portable alarms or speakers to monitor behavior at random moments. Records kept by a research staff that depend on direct observation of individual behavior are expensive and intrusive; they have been used in confined situations, such as schools, institutions, and workplaces.

Diaries, ledgers, and letters have been used in historical and biographical research, dispensing with representation; such *ad hoc* observations have been done in too many ways to list here. *Unobtrusive observations* is the beguiling name given to ways for counting diverse marks left unintentionally by humans for the purpose of measuring some of their behaviors [Webb et al. 1966]. This resembles the techniques of archeology and of anthropology for residuals from vanished people, such as inspecting their trash and bones.

3. *Longitudinal studies*, *follow-ups*, and *multiround surveys* are all names that have been used for repeated samples for observing populations over intervals. The repeated observations can have several purposes and diverse designs, and both purposes and designs are discussed in Section 6.2. Here we merely note that repeated studies are alternatives to retrospective data and to registers and records.

6.2 PURPOSES AND DESIGNS FOR PERIODIC SAMPLES

The title of this section emphasizes that designs should follow, not precede, purposes and follow them closely. Designs over time are generally made for regular periods, but they can serve repeated samples with irregular periods and continuous studies if those are feasible. *Samples* is a broad enough term to refer to experimental designs and observational studies, to which these concepts and designs can also be applied, though they originated chiefly from sample surveys.

In Table 6.2.1 we note five purposes and six designs. The first four are paired with similar letters on the same four lines. These pairings call attention to designs that best serve, with reduced variances, each of the four purposes. Most periodic studies have several purposes and thus we should face—not necessarily solve—the difficult problems of multipurpose designs (7.3). Actually, current levels (A) and net (mean, macro) changes (C) can be served with any of the six listed designs, but with some increase in the variances or in costs. But individual (gross, micro) changes (D) need panels, and cumulations (B) need some changes. Often reasonable compromises become possible—to the degree that purposes can be defined. Furthermore, extraneous considerations may rule out some designs (e.g., overlaps may be either prohibited or enforced) and thus force the use of less efficient, but still valid, designs. The chief variation in these designs concerns the amount (and kind) of overlaps between periods. The rotation scheme of complete overlaps shows, with *aaa–aaa*, that the periods have all common parts; the nonoverlap with *aaa–bbb* shows none; and the partial overlap *abc–cde–efg* shows *c* and *e* as one-third overlaps between succeeding periods only.

This section concentrates on the effects of varying proportions of overlaps P in diverse designs on different purposes; in complete overlaps $P = 1$, in nonoverlaps $P = 0$, and in partial overlaps $0 < P < 1$. The purposes are

TABLE 6.2.1. Purposes and Designs for Periodic Samples

Purposes	Designs	Rotation Scheme
A. Current levels	A. Partial overlaps $0 < P < 1$	*abc–cde–efg*
B. Cumulations	B. Nonoverlaps $P = 0$	*aaa–bbb–ccc*
C. Net changes (means)	C. Complete overlaps $P = 1$	*aaa–aaa–aaa*
D. Gross changes (individual)	D. Panels	Same elements
E. Multipurpose, time series	E. Combinations, SPD	
	F. Master frames	

discussed in terms of variances for estimated means, because means (and percentages, rates, proportions) are both the most used and the simplest estimates to treat. Effects on other estimates will not be entirely different but they are too many, diverse, and difficult to be explored here.

Effects on the variances of means from different proportions P can be treated clearly in this brief section. Other questions of biases, of feasibilities, of costs are often even more important, but also more difficult. They are treated elsewhere in this chapter, and in many places, including all sampling books [e.g., Kish 1965a, Sec. 12.4–12.6]. Much of the discussion also assumes for simplicity that the periodic samples are of the same size or of the same sampling fraction; but changes in sizes, fractions, and designs are possible, and even desirable in some cases, as noted below.

6.2A Current Levels and Partial Overlaps

Current levels is one name for the most common type of estimates for single "points" in time, whether the point of reference period is a single day or minute, or a week, a month, or even a year. *Static* estimates may be a better name than *current*, because the time of presentation and use may be years (or centuries) after the reference period, and *one-shot* or *snap-shot* have also been used in descriptions. "Cross section" has been used commonly to distinguish single-period surveys from longitudinal surveys. But that distinction perpetrates a confusion of temporal and spatial aspects, and it originates in financial constraints that confine broad cross sections to single periods and confine longitudinal studies to restricted sites. However, many studies of single periods have been confined to restricted sites; on the other hand, periodic studies of national cross sections, though costly, have emerged and yield valuable data (6.2D). Current levels for each period serve as important first results even of periodic studies.

Variances of current estimates are the same for complete overlaps $P = 1$ and nonoverlaps $P = 0$; they can be expressed briefly for means as $\text{Deft}^2 S^2/n$, where Deft^2 is the effect of the sample design on either the element variance S^2 or the sample size n (7.1).

That simple formula also holds for *simple* means from partial overlaps $(0 < P < 1)$. But statistics based on them can utilize the overlap P for a reduction of the variance with a complex mean: With help of the correlation R^2 between surveys within the sample overlap P, the portion $(1 - P)$ of the *preceding* sample is combined with the current mean to improve it. The variances are reduced by the factor $[1 - (1 - P)R^2]/[1 - (1 - P)^2R^2]$. This is a clever technical contribution much explored by sampling theory [Cochran 1977, Secs. 12.11–12.12].

The actual gains unfortunately tend to be modest in most practical situations; the maximal reduction in variance, utilizing optimal proportions P and optimal weights, is in the ratio $[1 + \sqrt{(1 - R^2)}]/2$. The reductions increase to about 33 percent only for very high R^2 values, seldom seen in practice; for $R = 0.9$, for example, $[1 + \sqrt{(1 - R^2)}]/2 = 0.72$; this ratio is obtained either with the optional $P = 0.30$ or with $P = 1/3$. For $R = 0.6$ that ratio becomes 0.9, only a 10 percent reduction of the variance. We note that overlaps of $P = 1/3$ or $1/4$ yield close to optimal reductions for most values of R^2, even when these vary greatly for diverse variables. This is a remarkably robust and useful result. Note also that in a long series the complex mean from the preceding sample can already benefit from reductions from its predecessors, and that using a longer series provides further slight reductions—only slight, unfortunately, because the factor $(1 - R^2)$ "decays" quickly with repetition; e.g., with $R = 0.9$ we never reach 0.60. Happily, for the other purposes of repeated surveys statistical theory is more productive as well as simpler.

6.2B Cumulations and Nonoverlaps

Cumulations refer to the purpose (and practice) of accumulating, pooling, and aggregating sample cases of individuals. No standard distinctions among these four terms exist and there is little literature. We can distinguish later (6.6A) between cumulating cases and combining statistics, but here both stages are treated with combined means, based jointly on periodic samples. With the greater availability of periodic samples their use is increasing; hence we have devoted Section 6.6 to those new uses.

Means based on several periodic samples covering a longer interval is the purpose we treat here, but the implications are similar for other statistics, such as regressions and other analytical statistics. The aims of cumulations are threefold (6.6B). First, they obtain greater precision, with lower variances from larger sample bases, especially important for smaller domains. Second, from the larger sample bases of cumulations we also expect greater spatial spreads of the design, so they can better cover small domains. Third, they can cover temporal variations—seasonal, cyclical, irregular—over longer intervals that include several periods.

Samples with no overlaps, $P = 0$, are best for cumulations. They are simpler and also yield lowest variances: $S_j^2/2$ for two periods and S_j^2/J for J periods, where the S_j^2 are variances for single periods assumed to include Deft^2 and factors like $(1 - f)$. For overlapping samples, however, the correlations R between periods increase those variances (6.6F). For example, the combined mean of two similar samples has $\text{Var}\,(\bar{x}_1 + \bar{x}_2)/2 = S^2(1 + 1 + 2PR)/4n = (S^2/2n)(1 + PR)$; this appears as the mirror image of $(1 - PR)$ for the difference of two means in (6.2C). Thus an overlap of $P = 1/3$, with $R = 0.75$ has a factor of 1.25 for increase of the variance. This can be decreased with better estimators and variance increases can be held under 50 percent while using the overlaps. As ratios of the variances these factors are much smaller than the threefold or even sixfold reduction possible for variances of changes. (See Table 6.2.2.)

Thus cumulations can be had even with partially overlapping samples; good compromises can be obtained, for example with $P = 1/3$, which is optimal for current levels and not bad for net changes. However, optimal allocations of $P = 0$ for cumulations remain in conflict with optimal $P = 1$ for measuring changes.

6.2C Net Changes of Means and Overlaps

Net change refers to the difference $d = \bar{x}_1 - \bar{x}_2$ of means between two periods; whereas *gross change* deals (in 6.2D) with the total changes of individuals, some of which remain hidden (because they cancel) in the net change of means. Measuring net changes are common and important aims of surveys and studies, and they are also related to other uses of the data. Perhaps the most common forms are differences in dichotomies, denoted by proportions $d = p_1 - p_2$, and in similar rates and ratios. We shall use the form $d = \bar{x} - \bar{y}$, which avoids subscripts and better symbolizes the more general applicability of the design and concepts developed here.

Net changes and differences $d = \bar{x} - \bar{y}$ denote aspects of design where, happily, statistics can yield great gains. The variance of $(\bar{x} - \bar{y})$ can be greatly reduced when the pair of variables have high positive correlations R in overlapping samples. Furthermore, we now turn to several aspects of great *flexibility* that may be explored in statistical designs for net differences.

1. The variances of mean differences are reduced by factors $(1 - R)$ in complete overlaps; this is the extreme (with $P = 1$) of the factors $(1 - PR)$

that may be obtained from partial overlaps. Hence for minimizing $\text{var}(\bar{x} - \bar{y})$ complete overlaps would be best. But partial overlaps are used in practice: (a) for reasons of feasibility, to reduce burdens, fatigue, and biases of respondents (6.2D and 6.4); and (b) to reduce variances of other statistics in multipurpose designs.

It is simple to think of the variances as $(2S^2/n)$ for differences between pairs of samples of size n without overlaps; $(2S^2/n)(1 - R)$ with complete overlaps, and $(2S^2/n)(1 - PR)$ with partial overlaps P. The S^2/n assumes simple random sampling and for complex samples the design effects Deft^2 should be included. But for differences and changes the factors Deft^2 tend to be smaller (closer to 1) than for the means (2.2, 7.1). Therefore reductions obtained from overlaps in complex samples, where Deft^2 are large for single means, may even be considerably greater than indicated by the factors $(1 - PR)$.

2. We may obtain almost the full reductions of complete overlaps even from partial overlaps by using improved estimators of the differences. These estimators are useful when circumstances may prevent complete overlaps but still permit partial overlaps. In those estimators the overlap portion P gets larger weights than the nonoverlap portion $1 - P = Q$, by the factor $1/(1 - R)$, because elements in the overlap contribute that much less to the variance. This improved estimator of the difference is

$$\hat{D}(\bar{y} - \bar{x}) = [P(\bar{y} - \bar{x})_p + Q(1 - R)(\bar{y} - \bar{x})_q]/(1 - QR). \quad (6.2.2a)$$

Its variance may be expressed, for two srs samples of size n, as:

$$\text{Var}[\hat{D}(\bar{y} - \bar{x})] = \frac{(1 - R)S^2}{(1 - QR)n}. \quad (6.2.2b)$$

These effects are shown in Table 6.2.2 with $a = (1 - PR)$ for the simple difference and $b = (1 - R)/(1 - QR)$ for the weighted difference, where $Q = 1 - P$.

The factor $(1 - R)/(1 - QR)$ approaches $(1 - R)$ for high values of R (where most important) and for higher values of P, say $P \geq 2/3$, as seen in Table 6.2.2, comparing the last two rows. High values of R are common for stable characteristics that can be well measured, but not for volatile, or poorly measured, characteristics or attitudes. Negative values of R must be rare, but that side of the table with negative values can be used to see what happens with sums of two means $(\bar{x} + \bar{y})$ when the factors are $(1 + PR)$ (6.2B). We also note again (as in paragraph 1 above) that the factors Deft^2 in complex samples may enhance considerably the gains from overlaps, because Deft^2 are less for the differences.

TABLE 6.2.2. Effects on Var($\bar{x} - \bar{y}$) of R_{xy} for Four Proportions of Overlaps P[a,b]

		Negative Values of R_{xy}					0	Positive Values of R_{xy}						
P		−1.0	−0.8	−0.6	−0.4	−0.2		0.2	0.4	0.6	0.8	0.9	0.95	1.0
1/3	a	1.33	1.27	1.20	1.13	1.07	1.00	0.93	0.87	0.80	0.73	0.70	0.68	0.67
	b	1.20	1.17	1.14	1.11	1.06	1.00	0.92	0.82	0.67	0.43	0.25	0.14	0
1/2	a	1.50	1.40	1.30	1.20	1.10	1.00	0.90	0.80	0.70	0.60	0.55	0.52	0.50
	b	1.33	1.29	1.23	1.17	1.09	1.00	0.89	0.75	0.57	0.33	0.18	0.10	0
2/3	a	1.67	1.53	1.40	1.27	1.13	1.00	0.87	0.73	0.60	0.47	0.40	0.37	0.33
	b	1.50	1.42	1.33	1.24	1.12	1.00	0.86	0.69	0.50	0.27	0.14	0.07	0
1.0		2.00	1.80	1.60	1.40	1.20	1.00	0.80	0.60	0.40	0.20	0.10	0.05	0

[a]From Kish 1965a, Table 12.4.III.

[b]Effects are $a = (1 - PR)$ for simple differences and $b = (1 - R)/[1 - (1 - P)R]$ for optimally weighted "composite" differences. Two equal, srs samples are assumed. Negative values of R_{xy} are rare, but these columns also show the effects of positive R_{xy} on $(\bar{x} + \bar{y})$.

3. Great flexibility can be used in choices of sampling units for the overlaps. It is necessary to use elements as sampling units for gross changes from panels and they generally yield the highest values of R, hence the lowest variances. But they also have great problems (6.2D, 6.4), and therefore in many situations larger units, clusters of elements, must be used instead for the overlapping units.

Compact area segments containing several dwellings and their occupants have been widely used for overlapping samples [USCB 1978; Kish 1965a, 9.5, 10.4, 12.5C]. Identification of dwellings and persons with the segments is feasible if well done. Each period's sample retains its character as a probability sample of the population, despite the moves of households, families, and individuals; despite births, deaths, and migration the stability of area segments remains representative of its inhabitants. It is true that, because of those changes and moves by the elements, the correlations R between periods are proportionately reduced; but the reduction affects only the changing portion. (This is currently around 18 percent yearly in the United States.) Hence overlaps based on segments retain most of the correlations R for measuring net changes.

4. The periods of overlapping can be chosen flexibly to reduce variances, especially for the comparisons that seem most needed. For example, *abc–Cde–Efg–Ghi* can represent 1/3 overlaps between succeeding quarters and yield good reductions of variances, for high values of R and with improved estimators (Tables 6.2.2). Note that half of the sample segments (c, e, g) is used twice (capitals show repeat appearances) but another half (b, d, f) only once. Thus 8 of the 12 segments are new each year and they contribute more to cumulations. We also learned (6.2A) that 1/3 overlaps are useful for estimating current levels. However, we may prefer cumulations over four quarters and also overlaps between years; then we can use: *abc–def–ghi–jkl–Amn–Dop–Gqr–Jst*, and so on. For other combinations see 6.2E [also Kish 1965a, 10.4, 12.5B].

5. In partial overlaps one may also vary the sizes of the nonoverlapping portions. While keeping the size of the overlap n_c the same for two or more surveys, the nonoverlapping portions $(n_x - n_c)$, $(n_y - n_c)$, etc., of successive periods may differ by design, so as to satisfy diverse purposes (6.5). For example, in one extreme $(n_x - n_c) = 0$, when one entire sample $n_x = n_c$ is added to another $(n_y - n_c)$ for a second period; or contrariwise a subsample of n_c of the first period may become the entire sample $n_y = n_c$ for the second period.

6. Greater flexibility may be used in the second and later waves of interviewing, or generally in the data collection in the field. The first wave must bear the initial costs of selection, contact, cooperation, and some basic, core information that later waves may reduce or omit. Therefore in later

TABLE 6.2.3. Effects of Diverse Overlaps on Variances of Differences of Means (Assumes $S_x^2 = S_y^2 = S^2$ and $\text{Deft}^2 = 1$)

Design Types	Sample Sizes for Specific Designs	Effects on S^2/n
A. Partial overlap	$n = n_x = n_y, n_c = Pn$	$2(1 - PR)$
B. Nonoverlap	$n = n_x = n_y, n_c = P = 0$	2
C. Complete overlap	$n = n_x = n_y = n_c, P = 1$	$2(1 - R)$
D. Subset	$n = n_x, n_y = n_c = Pn$	$(1/P + 1 - 2R)$

waves the costs per case (element, interview) can be made lower (a little or much) than on the first wave. Later waves may sometimes be done by different methods, perhaps by telephone or mail instead of personal interviews. This helps to explain the large overlap portions P on surveys, larger perhaps than are indicated by variances per case (n). Thus in the Current Population Surveys, overlaps of $P = 7/8$ are used, with the last seven of eight waves conducted mostly by telephone interviews [USCB 1978]. In some situations responses may also be better in later waves, but that is a complex and difficult subject; there is more on both costs and response in Section 6.4 on panels.

7. We avoid derivations in this book, but a simple and brief development of the variance $2S^2(1 - PR)/n$ for two overlapping samples ($0 \leq P \leq 1$) appears (7.2B). It is a simple sum of the variances and the covariance of two samples. It helps to explain the origin, meaning, limitations, and modifications of the simple factors in the last column of Table 6.2.3, which we used so often in this section.

6.2D Panels for Gross Changes of Individuals

Panels denote samples in which the same elements are measured on two or more occasions for the purpose of obtaining *individual* changes, $d_i = (x_{i2} - x_{i1})$. From a good sample of the d_i we can estimate the distribution of individual changes for the N elements in the population. Furthermore, from the mean of these *internal* changes of individuals we can also estimate the net, mean, external change: $\Sigma(x_{i2} - x_{i1})/n = \Sigma x_{i2}/n - \Sigma x_{i1}/n = (\bar{x}_2 - \bar{x}_1)$. However, from the net change of means one cannot estimate (directly) the gross change of individuals. This duality of changes has various names: individual/mean, gross/net, internal/external, micro/macro. Panels are sometimes called "strictly longitudinal studies" (6.1A).

Only panels can reveal the gross changes behind a net change; for example, a 2 percent net change of behavior may hide x percent canceling changes, where x may be small or large, unknown. Strong models could substitute for panel samples in theory, but in reality these exist only for some individual variables: age, parity (births) for women; some incurable, chronic diseases and infirmities; some acquired and permanent immunities; years of education; etc. Sometimes changes can be traced reliably from memory or from records. But often models and memory are both lacking and unreliable, and only panels can yield the data needed for individual, micro changes. These are needed not only for their frequency, but also for the dynamics of relationships and causation.

On the other hand, panels may be too difficult and not feasible for diverse reasons (mortality, mobility, refusals) discussed in Sections 6.3 and 6.4. Often, however, neither advantages nor disadvantages seem absolute, but they should all be weighed against each other. Here we need to clarify differences between panels and complete overlaps. Panels define special cases of complete overlaps when the sampling units are the elements themselves. But sampling units such as area segments used for overlaps differ from panels because of mobility and mortality in the population. Overlapping samples based on stable area segments can yield good current estimates and net changes; they have been so used in many surveys, e.g., the CPS [USCB 1978]. Area segments are more stable in rural portions, less in cities, and even less on their suburban fringes. Such stability (in degree and in time span) also describes their value for measuring changes [Kish 1965a, 9.5, 12.5C].

With their unique advantages panels are revealing results undiscovered by other methods, but they are not common because of the difficult problems associated with their use. Even less common are complete overlaps ($P = 1$) because they would have most of the problems without the completeness of panels. Since area segments have fair stability of people in short periods (about 82 percent over a year in the United States, 1985), the variance of mean (net) change ($\bar{x}_2 - \bar{x}_1$) is reduced by the factor ($1 - R'$), where R' is little less than R from panels. Other benefits (lower costs) and some, if any, disadvantages (i.e., refusals) are also proportionately inherited.

6.2E Multipurpose and Combined Periodic Designs

Most periodic surveys can, should, and do serve several purposes. Current estimates and net changes can be readily satisfied jointly using any proportion of overlaps, and applications are presented below in designs 1 and 2. For the other five designs described here (3, 4, 5, 6, 7) there is need, but no actual examples, I believe.

Cumulations of data can be readily added to either nonoverlaps or partial overlaps, but they require special design for spatial spreads (3, 4). The chief conflicts come from panels, because panels require special attention. They have two problems that conflict with two desirable properties of the other samples: first, possible loss of representativeness through attrition, and second, loss of cumulation. The compromises of these two conflicts make the combinations in designs 6 and 7 especially interesting; these are elaborated in Section 6.5.

1. *Partial overlaps* can be designed for both current levels and net changes. High overlaps are better for net changes, but low overlaps (e.g., $P = 1/3$) are better for current levels and they can be made to yield low variances with improved estimators (6.2A2 and Table 6.2.2). But high overlaps are often used because of the lower costs of later waves.

Time series are products of periodic samples but they seem difficult to specify as a specific purpose for a special design. Perhaps reducing variances for each periodic level and also for net changes seems most important, and for both of these reductions partial overlaps seem best. However, if moving averages are needed, then cumulations should also be considered to reduce their variances.

2. *Multiple partial overlaps* can be designed to meet the needs of time series for lower variances for specified intervals. The CPS design has $P = 7/8$ for successive months and $P = 1/2$ for yearly intervals [USCB 1978; Kish 1965a, 10.4]; however, it has $P = 0$ for most other intervals, such as two years. It is difficult to foresee and to design optimally and specifically for all the comparisons that will be needed. Perhaps a *permanent partial overlap* may be tried (see SPD in 6.5 below).

3. *Cumulative partial overlaps* refers to possible modifications where the nonoverlapping portions would be deliberately designed to cumulate to broader and better spatial spreads along with temporal cumulations. These then also represent modifications of the time × space matrix idea (6.1C) to partial overlaps. Spatial cumulations for small domains may become important objectives for periodic surveys.

4. *Cumulated reference periods* could provide more precise current levels, though at longer reporting intervals. These could be combined with overlaps between those intervals. But within the intervals the reference periods would be cumulated without overlaps and perhaps with increased spatial spread for lower variances. For example, monthly reference (and collection) periods with $P = 1/3$ between quarterly reporting periods can be represented by: |*aaa–bbb–ccc*|*Add–Bee–Cff*|*Dgg–Ehh–Fii*|, with each |quarter| including three monthly surveys, each with three thirds, like *aaa*. It would be possible, of course, to compute monthly levels, but the design optimizes for reporting

both the current levels and the net changes for quarterly data. Yearly overlaps can also be added [Dahlström, Jos and Wahlström 1973, Kish 1986].

5. *Panels spliced with overlaps* have been used and described in several surveys to solve conflict between these two designs. Complete overlaps of area segments were needed to assure representative samples of voters for both periods despite changes in the segments' populations (about 35 percent over two years). However, panels were also needed for data on gross changes in voters, and the movers were followed. Thus the design covered two samples in which about 65 percent was common to both [Hess 1984].

6. *Split-panel designs* (*SPD*) would incorporate two separate designs that have conflicting properties, advantages, and faults. A portion, say $P = 1/4$ or $1/3$, would be for a panel for individual changes; it would also provide overlaps and thus reduce variances for mean changes and for current levels, with *correlations* (R) *with all periods*. The other portions $(1 - P)$ would provide nonoverlapping samples to permit cumulation; hence they should have increased spread for cumulating periods. The two sample designs could be quite distinct to suit efficiently the needs of each. But the measurements would need to be similar to permit the combination of the two sets of results into single series of statistics (6.5).

7. *Rolling and panel designs* would extend the SPD above: the cumulation of the nonoverlapping portion would be designed to become a rolling sample census of all spatial domains of the population (6.5).

6.2F Master Frames and Master Samples

It would be difficult to imagine periodic samples selected without the prior existence or the creation of master frames. Needed also for nonoverlapping samples, they become even more important for cumulations, especially for samples "rolling" over the entire spatial extent. They can yield many kinds of auxiliary data for improving the periodic samples, the improvement depending greatly on specific situations. The nature of frames, the quality and quantity of information in them, their availability and utility vary too much for a useful summary here. The term may refer to a collection of maps and ancillary data for either the entire population or a large selection of primary sampling units; these serve as frames for selecting needed samples. Or with further work a large sample (lists of segments or households) may be selected in an initial first phase, and then the actual samples for each new survey may be selected as needed from those lists already prepared. A "master sample" may actually contain personal data obtained in first-phase interviews, to be subselected for second-phase surveys [Kish 1965a, 12.6A; Wright and Tsao, 1983].

6.3 CHANGING AND MOBILE POPULATIONS

6.3A Four Competing Sources of Changes Over Time

Whenever we want to separate the factors (predictors) responsible for changes in effects (predictands), we encounter possible or actual disturbances, "noises," from a bewildering variety of other factors. Several of these factors also may be affecting the predictands, some accentuating the effects and some masking them perhaps. There exist so many possible factors that to list all of them appears unfeasible. I believe, however, that we may find the following four potential sources (kinds) of change useful for bringing some order to the multitude of possible factors. (See also the biases T1 to T4 in 3.5.)

For any study some factor from one of these sources will appear as the principal objective, and the other three sources become disturbing factors, potential or actual. When any disturbing sources, or factors therein, seem important, they must be brought under control with statistical methods, either in the design, in the analysis, or with field or office procedures. This separation of explanatory factors from disturbing sources is useful to the extent that researchers can apply it to their own research situations. We may add that in two separate analyses of the same set of data, first one source then another may become the objectives and predictors; with each effort the residual three sources become the disturbing sources to be controlled.

Any profound treatment would lead us into the methods and philosophy of some specific substantive field of research, and I cannot provide a good bibliography because the references are scattered across many fields. On the other hand, this brief list of sources can serve to remind us of the disturbing factors so that the statistical design can be shaped to control them. See also another classification from a different aspect in Section 6.6C.

A1. *Internal* sources refer to changes *within* individuals that occur *naturally* and usually *gradually*. Aging, growing, and learning are examples of changes that tend to come gradually without outside intervention. These processes are usually related to each other, but not at all perfectly, and we need to deal with each of them separately. Aging refers to processes that are automatic, unpreventable, and irreversible, and they take different forms for diverse elements to which we may want to generalize. Growing refers to changes in size and shape of elements; it is quite different, not at all automatic or irreversible, and may be negative. Growing is not entirely autonomous and internal, and may be affected by external influences (A2), but still it refers to internal effects. Learning deserves a separate name, though related to growth in a general sense; it is affected by external factors (A2), and possibly by the treatments (A4) through experimental contamination, or even by the procedures of measurements (A3). Negative forms of learning can be the effects of the study itself; they may be called tiring and fatigue or resistance and refusals. These are well-known dangers of repeated, longitudinal measurements, and they are affected by the measurement (A3) and the treatments (A4).

A2. *External or environmental* sources affect the elements from the outside, affect all or many of them *jointly* and often somewhat *similarly*; the effects on the elements are often relatively *abrupt* and *simultaneous*. Historical events come to mind, and usually disasters like wars or epidemics; but abrupt news may also be good, like a war's end or discovery of a new cure. External events may also be gradual and show other confusions with internal sources, but the broad distinction seems useful for hunting and listing disturbing factors. Let us be reminded again that an external event may be either the principal study objective or a disturbing factor to be controlled.

The *age–period–cohort* confusion has received attention recently in demography [Fienberg and Mason 1979, Rodgers 1982]. Cohort analysis follows a population (either in aggregate statistics or individually) of the same age, as its age increases regularly with successive periods. The contrast of age and period effects may represent internal (A1) versus external (A2) factors.

A3. *Instrumental change* in the procedures of measurement or observation may be a source of disturbing factors that need to be controlled, prevented, or reduced. Unplanned changes or drift in the measurement procedures may be a disturbing factor that can cause annoying surprise or (even worse) remain undetected. However, sometimes instrumental changes may represent the principal objectives of a study for comparing different methods or instruments of measurement between two (or more) waves, perhaps over the same population; and then the other three sources can present disturbing factors. Instrumental changes usually occur in the sample only, and this may also be true of A4; whereas the internal and external changes noted in A1 and A2 affect the population.

A4. *Deliberate interventions* and new treatments are likely to be the principal objectives of evaluation studies (3.7). They may originate in new laws; in innovations, inventions, or changes; or in experimental treatments. However, they may also appear as disturbing factors in studies of growth and learning, for example. Confusions with other external sources (A2) may arise, hence the distinction seems practically useful.

6.3B Mobility and Changes of Populations

The topic of mobility and changes of populations, like the preceding one, needs this systematic, joint treatment to avoid confusion (common for panels and for partial overlaps) between various sources of change in samples and in populations. These problems receive occasional and separate treatments, but a joint heuristic basis may reveal common features, which occur because the elements of the populations can move, appear and disappear, or change between (or during) the periods of the studies. Mobility and changes can be common and frequent for persons (and animals); even more so for families, groups, and institutions; but less so for dwellings and for geographical and geological features. The diverse kinds of mobility across boundaries of populations, as well as of sampling units, cause problems of location, definition, and identification for methods based on area frames. Other frames also can have analogous problems: People move into or out of schools, firms, institutions, groups, etc.

B1. *Three Components of Population Change: Changed Boundaries, Migration, Changing Elements.* These three components of change interact over time in the definition of the population—of a city, institution, organization, etc. Suppose, for example, that we study behavioral changes in the inhabitants of a city between two samples or censuses 10 years apart. In that period the city has expanded its legal limits, and, perhaps even more, its pattern of settlement. Thus four alternative limits—old or new, legal or social—can be used for defining the population limits. Meanwhile large immigrations of different social and ethnic groups could have occurred. The movements of boundaries and of people are disturbing factors for studies of changes within populations (which could be already complicated by age–period–cohort problems). For studies of any of the three components the other two appear as disturbing factors (6.6C).

Analogous problems of definitions appear in other populations. Changed national boundaries can interfere with historical comparisons, but they are less permeable and ephemeral than internal boundaries. Boundaries and limits other than geographical can be faced with analogous problems of definition; money values (e.g., income classes) need adjustments to some common "real" base; definitions of social standards (e.g., poverty level) often need realignments.

B2. *Changes Within Complex Elements.* Identification of elements, can become difficult when these are complex units, such as groups, organizations. Persons remain identifiable over periods—with the aid of biological homeostasis; but identification of families is complicated by marriage, divorce, migration, births, and deaths of its individual members [Duncan 1984]. Studies over time of firms, organizations, and institutions are difficult (though not impossible) but worthwhile. Their forms of homeostasis differ from that of persons. On the other hand, studies confined to their nonchanging portions would yield biased pictures of the whole population.

B3. *Populations of Events.* A probability selection of persons may be made too difficult in some situations by their mobility; sometimes, however, a solution may be found by redefining the problem and the statistics in terms of a population of events instead. For example, studies of customers of stores, users of libraries, patients of doctors, clients of lawyers, shareholders of companies have been substituted for samples of persons behind those events. The estimates can account for the knowledge that several events could belong to any person. Not only persons but other mobile biological, physical, or other elements may be similarly redefined in terms of events, appearances. For samples of airplane passengers the "legs" (stops) of flights may be the events to be counted.

If the number of events (e_i) can be obtained for each person (element) in the sample, by weighting with $1/e_i$ the statistics can be recomputed for a population of persons [Kish 1965a, 11.2C]. (For related problems see 7.4B, 7.5.)

B4. *Conflicts of Location Versus Allocation.* This is an attempt to cover with heuristic unity a variety of problems that arise because the location of mobile elements at the time of collection (of a census or sample) differs from a desired (proper) allocation.

a. *Dwellings* of usual residence are commonly used by censuses and sample surveys for unique association of persons and of families. They have been found to be remarkably useful for women in diverse cultures throughout the world by the World Fertility Surveys [Verma et al. 1980]. Nevertheless serious problems remain and their solutions, often imperfect, depend on specific situations and resources. They are often discussed [e.g., Kish 1965a, 9.1], and here we only note some of the most frequent problems. *Persons away* from their residences at collection time pose problems. *Callbacks* may be expensive and delay completion of collection. Data from others, from *proxies*, may be poor, or not acceptable (for attitudes). Collecting data from persons who are found distant from their dwellings raises problems of identification and duplication. *Travelers, tourists, and vacationers* represent three types of populations to be studied in motion, away from their dwellings. These studies may be sometimes better accomplished with a redefinition to a population of events.

b. *Nonhousehold populations*, i.e., persons not in dwellings, take diverse forms and receive diverse treatments. They may be in institutions like schools, hospitals, prisons, or military installations. They may live in transient hotels, boarding houses, or camps; in trailers, mobile homes, or autos; or on streets or nowhere. They do not form a random population subset, but tend to be male and young, often poor but sometimes rich. They may be omitted in small samples, or covered with special methods in large samples and in censuses.

c. *Nomads and migrants* constitute specific populations and large, special problems in some situations. Nomads generally move in families or tribes and often in some regular, cyclical, or seasonal manner, but often in diverse patterns. Migrant workers often move as individuals, sometimes in families, seldom in large groups; but often in large numbers and in irregular and shifting ways. References exist on surveys and censuses of nomads [UN 1977].

d. *De facto and de jure allocation* are terms given by census offices facing a dilemma between two alternative procedures: *de facto* for locating persons at the actual sites of enumeration and *de jure* for locating them at their usual residences. Each has problems and these vary with situations and resources and they differ for diverse populations. *Institutionalized* populations (such as students in dormitories) may be counted *de facto* at the institutions (universities) or *de jure* at their home residences. *Hospitals* are the actual sites of births, deaths, and illnesses, but allocating the patients to them would result in distorted vital and health statistics. *Travelers and vacationers* may be noted here again as problems of allocation. *Working or daytime populations* of central cities, and of other specialized worksites (mines, harbors, factories), pose problems for service statistics, when residences and workplaces are separated.

e. *Exclusive units for unique appearances* can be used for locating persons and other mobile units. A simple version of this is the census procedure of using a single reference date for uniquely locating a person moving during the collection interval. A more complex version may be needed for mobile elements such as nomads and migrants and for wildlife and other mobile elements. Suppose the elements must be allocated *de facto* to the site where found at the time of enumeration; suppose also that the enumeration must be spread over a longer interval; suppose also that the elements move, migrate in an unpredictable and probably not strictly random manner. It is still possible to obtain a sample in which every element has a known expectation of appearance, hence unbiased estimates, although the appearance becomes a random variable. We need to assume that at any one time the element can be identified in one and only one unit at any time during the collection interval and that a probability sample of these units (often area segments) is selected.

If the overall selection probability of the units is f, the expected appearance of any element is also f; hence dividing by f produces unbiased population totals. Any element may appear $0, 1, 2, \ldots, k$ times, where k is the number of "times" in the total interval. The time may be a day (for nomads) or shorter. Duplicate appearances are accepted, but they are rare, except for very large sampling fractions and very mobile populations, i.e., large k over the interval. With $f = 1$ we would have a "census" but of a variable size.

6.4 PANEL EFFECTS

Panel studies possess dual aspects and functions. On the one hand, a panel study of k periods can be compared with k distinct samples, and this comparison is more basic (6.2). Thus a sample of n elements (households or persons) may be observed in k periods for a panel, compared with a total of kn elements in k nonoverlapping periodic samples of n each. The costs per interview are somewhat cheaper for a panel. This comparison of relative costs of the two sampling methods for obtaining "static" information in k periodic samples is seldom made, but outlined in 6.4A [Freedman, Thornton, Camburn 1980; Duncan 1984].

On the other hand, there are interesting comparisons, chiefly in epidemiology, of *retrospective* studies versus *prospective* studies, each confined to one sample. In retrospective studies memory and records are used to retrieve the needed longitudinal information. These comparisons are outlined in 6.4B, and they differ greatly in assumptions and in costs from the first comparison in 6.4A. But the costs per element are much higher for the prospective panels than for retrospective studies of only n elements.

6.4A Panels Versus Distinct Samples

1. *Initial self-selections.* Any sample of humans probably involves some form and some amount of volunteering, hence self-selection, hence potential bias in representation. These are all motivational factors and their effects vary so widely that no quantitative guidelines seem possible. However, it has been noted often that the rate of refusals is increased considerably when respondents are asked for cooperation in a long-continued panel after the first call.

2. *Attrition* continues after the first call, but at a much reduced rate. This attrition has two forms: refusals due to "panel fatigue" and nonresponses due to disappearances that cannot be traced. We distinguish these from losses due to temporary nonresponse, to mortality, to changes, and to mobility, all of which are treated separately. The refusal at the first call may

be, let us say, as high as 20 percent, but the attrition after that may be as low as 1 or 2 percent on each call. Nevertheless, these small losses can also accumulate to a sizable total after many periods. But these effects are extremely variable; and fatigue and refusals are much less in rural and in less developed areas (see 14 below).

3. *Temporary nonresponse*, either not-at-home or refusal, may be considerably higher than attrition; say 3–6 percent versus 1–2 percent, depending greatly on timing and kind of procedures. Hence they must be included in later calls, and their data interpolated with retrospection and with imputation.

4. *Mobility* must be treated distinctly from inevitable attrition. First, mobility may be much greater than attrition, depending on the population and on the time interval covered; hence losses could accumulate to prohibitive levels. Second, they can be reduced or eliminated (almost) with enough care, effort, ingenuity; and the literature conveys much good advice, specific to the situation but translatable to others. Mobility has entirely different effects on panels than on overlapping sampling units, such as area segments, which are self-correcting and reflect (in expectation) the changing population. Longitudinal studies of restricted sites with permeable borders, such as a single area, will reflect great mobility (3.1A).

5. *Changes* of the elements can be considerable for complex units, like families, groups, organizations, institutions, firms. Dealing with them in a panel requires much skill, knowledge, and experience. Such changes generally reflect similar changes in the population, as does mortality.

6. *Mortality* affects the entire population, and its treatment would be different and simpler for a study defined by and confined to the initial sample and population. Within that definition the panels suffer no special defects, as compared with changing samples, either from mortality, from other forms of outmigration, or from changing elements (6.3). That is why we separate the panel effects of changes, mortality, and births (5, 6, and 7), from attrition and other specific defects of panels (1, 2, 3, 4).

7. *Births and immigration* should be introduced into longitudinal studies defined by ever-changing, living, and complete populations, with births into as well as deaths from them. They must include some method for introducing births and migration, in contrast to studies confined to the initial population (6 above) and in contrast with nonpanel methods defined by stable units, such as area segments, even in overlapping samples.

8. *Retest reactivity*, panel bias, panel contamination, sensitizing, or learning are all names given to the fear that the experience of past interviews (observations, enumerations) and the anticipation of future ones will change the behavior and attitudes, opinions of the individuals in the sample. I cannot say anything useful on this deep and controversial topic, except that

there is little information on effects in contrast with widespread extravagant fears. Any effects would depend on many factors involving the nature and timing of observations, of the study variables, and of the population. (See item 14 for positive effects.)

9. *Reinterview laxity* has been raised as a possible source of bias: that both the respondent and the interviewer may be subject to inertia and to similarity of the answers, which probably they must (unintentionally) recollect from past interviews. Interviewers may also become somewhat less careful generally on return visits. The "rotating group bias" of the Current Population Surveys is the best-known example, though a rather confusing problem [USCB 1978]. On the contrary, we also present (in item 14) arguments in favor of the familiarity gained in panels.

10. *Checks and controls* are desirable to guard against possible biases from the use of panels. These can take so many forms and are so dependent on specific situations that listing them here seems futile. Checks can generally be of two kinds: comparisons with available background variables, such as age and sex, and of the study variables that are more critical but also more difficult to validate. Perhaps best would be a comparison of each wave with a new changing sample; this is a feature of Split Panel Designs (6.5). However, it would be wrong to assume that any difference between panels and new samples measures biases in panels, only because we are less familiar with them! (See items 12, 13 and 14.)

Methods for controls and corrections can be analytical with some form of weighting or with imputation, which may be easier, but they can complicate later analysis. It would be more difficult to arrange, but also ultimately more useful to "correct" the sample for panel losses with supplements of the missing types; but such supplements of new members will lack the desired history of the panel cases.

Fortunately, after those 10 possible defects we come to four possible and considerable benefits of panels; three of these advantages (12, 13, 14) are also shared in good portion by overlapping sampling units.

11. Only *panels* yield data on individual changes, as discussed in 6.2D.

12. *Lower costs* per interview than those for changing, nonoverlapping samples are common, in spite of the widespread hostility to panels. First, only the first wave bears the sampling costs, both in the office selection and designation and in the field work of identification and gaining access and cooperation. Second, the basic background data concerning individuals ("face sheet data") are borne mostly or are more costly in the first wave. Third, acquaintance with the unit (household) facilitates contact; for example, the timing of calls (interviews). Fourth (and this is most variable), later calls can cost *much less*, if done by telephone, mail, or some other cheaper procedure, on all or on most of the sample, when this seems not

feasible for the first wave. (This in my view is the chief, though largely neglected, reason for the large overlaps in some current labor force surveys.)

13. *Errors removed or reduced* can be a considerable advantage of panels, if procedures are introduced to check for differences and for consistency.

14. *Familiarity* with the sampling units and with the individuals can often have positive results, in contrast to the negative and feared effects of attrition (2), reactivity (8), and laxity (9). For demographic surveys in developing countries it has been noted that,

> The survey staff will master their duties better and learn to know the sample areas and even the population. For their part the respondents, meeting interviewers they already know, become more relaxed and willing to answer questions. It has been reported in several surveys which have lasted three or more years that initial suspicion and reserve have with repeat visits given way to trust and the interviewers have been received with pleasure as old friends (Cantrelle 1974; Nepal 1976; Iran 1978). [Kannisto 1983; UN 1984]

6.4B Prospective Panels Versus Retrospective Studies

This contrast differs greatly from the contrast in 6.4A of a panel with a similar total number of visits. Here instead, the use, value, and cost of a panel of several waves are contrasted with a one-wave study, which depends on retrospective recollection of data over time. This contrast is best developed in the literature of epidemiology and public health as "retrospective" versus "prospective" studies of diseases and risks [Cornfield 1956; Cornfield and Haenzel 1960; Berkson and Elveback 1960; Greenberg 1969]. Other researchers can profit from these and from a rich source of further references therein.

The two kinds of contrasts give extremely different views. This is especially true of costs, because panels seem less costly per interview than a similar number of new waves, but prospective panels are much more costly per individual than a one-time retrospective study. Between these two extremes of contrasts other comparisons and compromises are possible and my two listings of panel effects may help the readers to fit their own situations. We may also think of a prospective panel of several waves and each collecting not only current data, but also retrospective data; we may also think of panels newly started with each wave and then followed prospectively. These models yield rich data, but are expensive [Bachman and Johnston 1978]. We now turn to a listing of the problems of prospective panels that lead often to using retrospective studies (1, 2, 3, 4), followed by the doubts inherent in their use (5, 6, 7, 8).

1. *Higher costs.* Panels of several waves are bound to incur considerably higher costs than a retrospective study of single observations (interviewers) on a similar number of individuals. A simple model for panel costs per individual would be $C + wf$, where C refers to contact costs of selection, identification, acquaintance, and basic background data (6.4A.12) and f refers to follow-up costs for each of the w waves. For a retrospective study the cost per individual is only $C + r$, let us say, where $r > f$, for a longer interview. We may say that $C + r < C + wf < w(C + r)$ and that retrospective studies cost less than panels, especially as the number of waves w increase, though not by the factor w.

In a situation with a small number w of waves, and when the contact costs C are larger than f, the cost difference may not be overwhelming. Furthermore, the size of the panel may be decreased by a factor d; then $C + r = (C + wf)/d$ describes the decrease d in relative sample size needed for a panel of w waves, to bring it down to the cost of a retrospective study.

2. *Rare events.* Prospective studies of panels with several (many) waves can become especially expensive when the proportions of "susceptibles," (X/T), or of "diseased" (D/T), and especially of "susceptibles with disease" (XD/T) are small. (See Figure 6.4.1 for symbols.) This common situation would require following a large total sample T chiefly to find eventually a small proportion XD/X among the susceptibles (who may themselves be infrequent) and probably a smaller proportion OD/O among the resistants.

This situation leads to the use of retrospective studies where D diseased individuals and H healthy are found first, and then through investigation both are classified into susceptibles and resistants: the D into XD and OD, the H into XH and OH. Thus in retrospective studies we find cases with effects (responses) and then investigate them retrospectively for causal factors (stimuli). This retrospective, backward path to causation is subject to the criticism and to the cautions referenced earlier.

When searching for causal factors, the most desired statistic may be the ratio $p_x/p_0 = (XD/X)/(OD/O)$. For example, the ratio of lung cancer rates among cigarette smokers is about 15 times that rate among nonsmokers. Since p_0 in the denominator should not be so small as to induce instability in the ratio, it may be wiser to use p_0/p_x when p_0 is too small and less stable than p_x (4.7).

When, however, we would estimate the magnitude of explanatory effects, it may be better to use the difference $d = p_x - p_0 = (XD/X) - (OD/O)$. For example, the difference in death rates from hypertension between cigarette smokers and nonsmokers shows more clearly the importance of that effect among all death rates than the ratio p_x/p_0. The magnitude of effects in numbers of deaths will be $N_x d$, with the difference in rates d "explained" by cigarettes multiplied by the number of smokers N_x.

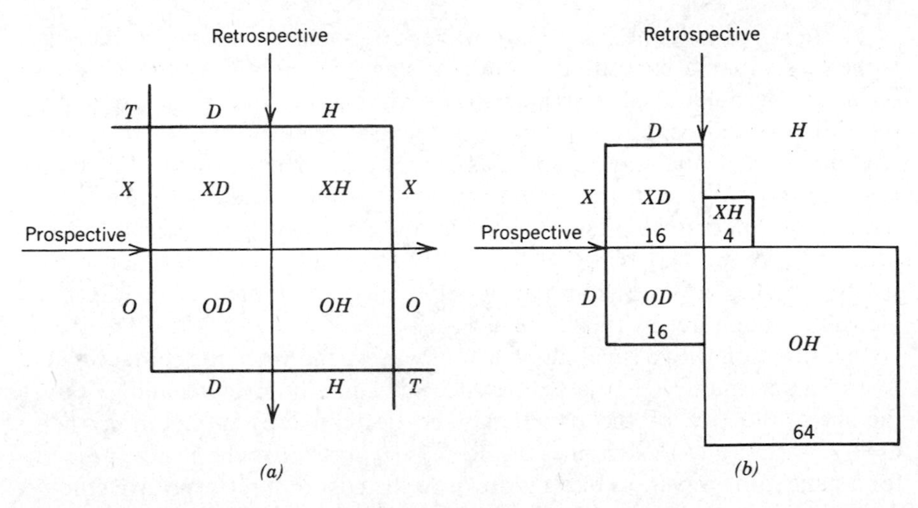

Figure 6.4.1. Prospective versus retrospective studies.

a. In prospective studies researchers *follow* two samples X and O to divide them into $X = XD + XH$ and $O = OD + OH$. The X/O distinction may denote positive/negative screening, $+/-$, or treatment/control, E/O, susceptible/resistant, etc. The D/H split may denote diseased/healthy, or affected/unaffected; $p/(1 - p)$, with symptom/symptom-free, etc. In medical screenings the XH are called "false positives" and the OD "false negatives." In *retrospective* studies the researcher finds two samples D and H, and divides these into $D = XD + OD$ and $H = XH + OH$ (on the basis of retrospective interviews, tests, diagnoses, etc.). In *surveys* the entire sample of T is split into the four cells and the analysis may go either way. Note that the "odds ratios" of prospective and retrospective studies are numerically equal, because $(XD/XH)/(OD/OH) = (XD \times OH)/(XH \times OD) = (XD/OD)/(XH/OH)$. However, such equality does *not* hold generally for the differences of those ratios, $(XD/XH) - (OD/OH) \neq (XD/OD) - (XH/OH)$ nor for the proportions $(XD/X) - (OD/O) \neq (XD/D) - (XH/H)$. In addition to those numerical problems, often grave differences also exist between prospective and retrospective studies and surveys in representation, selection, and response.

b. This diagram portrays a common situation: the positives are in minority $X < D$, and among them the "false positives" are fewer $XH < XD$; and "false negatives" are also a minority $OD < OH$. The size of the association shown is uncommonly strong, with the odds ratio $(XD/XH)/(OD/OH) = (16/4)/(16/64) = (4/1)/(1/4) = 16$; lung cancer rates for cigarettes/nonsmokers are about this strong. Note that the weakest link is the $XH = 4$ "false positives" in the denominator: If this were based on four cases out of $N = 100$, the ratio would be most unstable. The difference in the two prospective proportions is $.80 - .20 = .60$ and for retrospective $.50 - .07 = .43$.

The simple dichotomies can be extended to polytomies and to continuous variables. Furthermore, several explanatory variables, and especially controls for disturbing variables, may be introduced.

3. *Delayed results.* This may often be the principal reason for using retrospective studies, instead of waiting for years or decades, which the full unfolding of a prospective panel would require. It may be possible sometimes

to do a retrospective study soon and then to begin a prospective long-range project to allay eventually the doubts from the former (6, 7, 8).

4. *Panel fatigue, bias, mortality.* These problems have been treated in detail in 6.4A, but we may note that mortality and other selection biases are likely to be greater in retrospective than in prospective studies (7).

5. *Lack of randomization.* Random assignment of treatments seldom seems feasible, and often probability selection of individuals is also too expensive. However, it would be unreasonable to hold these imperfections against prospective panels, since these factors are likely to face greater hazards and doubts in retrospective studies.

6. *Biases of memory, recall, retrieval.* These cover the principal objections to retrospective studies, and much is written about them (6.1C.1).

7. *Mortality biases.* These refer to biases in the population arising from possible differential mortality (and other losses from attrition) between affected (D) and unaffected (H) individuals, within both causal classes (X and O). These biases may have greater effects in retrospective studies, because they may be traced in prospective panels.

8. *Selection biases.* In addition to biases in the population, retrospective studies may suffer more from selection biases. To find affected (D) cases when these are rare and to find small classes of susceptibles (X), probability sampling may seem too expensive, and judgment sampling may be used. Statistical inferences will be subject to doubts, though biases in the ratios and differences will be less serious than in first-order results.

6.5 SPLIT-PANEL DESIGNS

With the name *split-panel designs* (SPD) I hope to distinguish this new type of design from older designs with which it could be confused. It is not one of the four basic designs (panels, overlaps, nonoverlaps, partial overlaps) of Section 6.2. SPD includes a combination that performs all four purposes and some new functions in addition. Hence the new name I gave to this new type of design should help to avoid confusing it with older and better-known designs.

1. The basic notion is simply to add to a panel p a parallel series of nonoverlapping samples, denoted as *a-b-c-d*-etc. Thus the periodic SPD may be symbolized as *pa-pb-pc-pd.* The panel p yields individual changes, the nonoverlaps *a-b-c-d* can be cumulated into larger samples; and the combined sample provides the partial overlaps best for current estimates and for net changes. This combined use is the main feature of SPD; methodological comparisons and checks of the panel against the changing samples are side benefits.

Nevertheless, the sample designs for *p* and for *a-b-c-d-* can be separate and distinct, each "optimized" for its own uses; this does not prevent their combined use as one sample. But the populations covered and the measurements used must be closely similar for the combinations.

2. The differences from the classical, symmetrical methods of partial overlaps and the advantages of SPD over them deserve primary emphasis. The classical designs of partial overlaps have the elegance and appeal (usual and often misleading) of symmetry: Each sample group is in the sample for a similar pattern of waves. (The simplest is *ab-bc-cd-de*. . . .) No individual becomes any more fatigued than any other—but also no less. For measuring changes it provides the gains (often considerable) for those intervals that have been prespecified in the design: for example, monthly changes or yearly changes. But only for those!

However, my reading of actual situations tells me that critical comparisons will probably be discovered for later waves that differ from the intervals specified at the time of the original design. SPD has overlaps of *p* for *all* intervals between waves, and this feature gives SPD considerable advantage over classical, symmetrical designs for measuring net changes. Another advantage of SPD over partial overlaps of sampling units comes from the higher correlations of the same elements when these are subject to great mobility between units.

3. For measuring current (static) levels also, in the gains from partial overlaps, SPD has some advantages over classical symmetrical designs, though not great, clear, and uniform. Briefly, in symmetrical designs the prespecified overlaps (which may be designed primarily for measuring net changes) may not have the interval with the highest correlations, R; for example, the overlaps may be yearly, but highest R's may come monthly. On the other hand, SPD has overlaps for all waves. Furthermore, the weights may be "optimized" for maximal R for each variate.

4. The chief difference and the principal advantage of SPD comes from having a proper panel that classical rotations lack. Panels are necessary for measuring individual changes, but they also have unique problems (6.2D). Hence they may be called either indispensable or impossible. But SPD also has a great advantage over a complete panel: It allows for cumulated evidence to check against and perhaps correct the biases to which panels may be subject (6.2D, 6.4A). Similar checks can also be made in partially overlapping samples. Another advantage of SPD over ordinary panels comes from the possibility of using the changing samples to recruit replacements for panel mortality and for panel renewals.

5. If a "permanent" panel is not feasible or desirable it may be modified in several ways and still retain some or most of the advantages. For example, a modest and slow rotation may be built into it so that most of the variance

reductions from correlations are retained, and so as to minimize the loss of panel information. Perhaps as measures against panel fatigue or deterioration it may be possible to introduce complicated designs so that only parts of the panel appear in each wave.

6. If panels of individuals, following moving individuals, are ruled to be not feasible or inadmissible, most of the gains enumerated previously can still be retained, with the portion p becoming a complete overlap of sampling units, for example, of area segments. Some of the correlation may be lost due to mobile elements.

7. The relative sizes of p against the *a-b-c-d* ... portions depend on feasibilities and costs. In favor of a large p are the valuable data on individual changes, greater precision for net (mean) changes, and perhaps lower cost per interview, especially with telephone interviews. At one extreme only small changing samples would be retained to serve for methodological checks and controls and for recruiting replacements.

On the other hand, low values of p (about 1/3) are better for current (static) levels; also to reduce panel biases and problems. The gains in variances in complex estimates for current levels and for net changes (6.2A and C) are not very sensitive, and portions of p between 1/4 and 1/2 are all fairly good; that is, the *a-b-c-d* . . . changing portions may be anywhere from p to $3p$. Furthermore, for cumulating cases we want to increase the new cases, the sizes of *a-b-c-d* . . ., and hence to decrease the relative size of the panel p.

8. Thus another advantage of SPD may be its flexibility in the sizes of *a-b-c-d* . . ., the changing samples, which are fixed in rigid, symmetrical, classical designs. In SPD these sizes can be varied from wave to wave in order to fit changing budgets and changing needs. Of course, such changes would raise weighting problems for cumulated estimates, but they can be solved (6.6E).

6.6 CUMULATING CASES AND COMBINING STATISTICS FROM SAMPLES

6.6A Stages for Pooling

Combinations can take place at any of four successive stages of preparation, and the title aims to distinguish the first two. Methods will be discussed later for these first two, perhaps "pooling" or "aggregating" can refer jointly to both; but for numbers 3 and 4 only references to the literature are given here. Data from repeated samples for the same or similar variables and populations are becoming more common and available, hence methods for combining or cumulating those data are becoming of wider interest.

1. *Cumulating cases* can refer to aggregating, summing, amassing

individual elements from repeated surveys. This has been practiced without much theory on periodic surveys, and the 52 weekly samples of 1000 households each have been designed for cumulation by the National Center for Health Statistics [NCHS 1958]. Other examples are rolling samples and multiround samples mentioned elsewhere. However, individual cases can be and have been cumulated also from samples that have been designed and collected quite separately from each other.

2. *Combining statistics* from published results can be done for surveys and for experiments carried out in diverse places and at different times. Means, rates, percentages, totals have been averaged. Analysis of the Results of a Series of Experiments [Cochran and Cox 1950] is an early reference for combining experiments. Combining statistics from surveys is discussed later.

3. *Combinations of probabilities* from a group of related experiments can be done by a method due to R A Fisher. The probabilities P_i from k separate and independent tests of the same hypothesis, with each of the k values of P_i based on a 2×2 chi-square test, can be combined into $u = -2\Sigma \log_e P_i$, which has a chi-square distribution with $2k$ degrees of freedom [Anderson and Bancroft 1952, 12.6].

4. *Meta-analysis of research* is the name of a field of methods, originated recently [Glass 1976] and mostly in the fields of education, evaluations, and social psychology and not much in statistics until recently [Hedges and Olkin 1986]. The methods concern combining disparate research results on some single substantive hypothesis or theory, and they seem to range from a systematic qualitative review of research all the way to combining statistical results. They deal mostly and in detail with problems of measurement, concepts and theory, but less with statistics. A review of 36 articles by R J Light [1983] provides many more references. Summaries of research have been done before [e.g., Gilbert, Light, and Mosteller 1975].

6.6B Purposes for Cumulations and Combinations

I distinguish three general reasons either for cumulating cases or for combining statistics, but the first purpose seems more suited for cumulating. For other purposes for repeated surveys we refer to Section 6.2.

1. *Larger samples for domains*, especially for small domains and even for rare cases, are needed, both for proper domains (like areas) and for crossclasses (see 2.3 for these types). These needs occur in all national samples and other large-scale research; sample sizes are limited, whereas interest in details seems unlimited. On the other hand, specific designs for meeting these needs with periodic surveys are few, though, one hopes, increasing. The yearly cumulations of the weekly samples of 1000 households each in the National Health Survey [NCHS 1958] may be the best example.

Larger bases for domains have motivated the pooling of polling data [Miller 1983]: "British academics . . . have made increasing use of cumulated media poll data to get access to large samples even if they had to be second-hand samples" [Bonham 1954; Butler and Stokes 1974; Harrop 1980; Rose 1980].

Rolling samples for all small administrative units would be an extreme case of small domains; only a sample of the units (like counties or Enumerative Districts) would appear in any one periodic (e.g., monthly) sample, but the design would cover the whole population of such units (and perforce their composites) in some designed period (e.g., 60 months in 5 years) [Kish 1981].

2. *Averaging or aggregating over space* and over other domains occurs in all samples, but it seems heuristic to recognize this explicitly. All national samples, for example, yield averages or aggregates over domains that may differ a great deal. In some countries (India, China, Yugoslavia, Brazil) national samples may be composites of separate state samples. Extreme examples consist of combining the national results of the World Fertility Surveys into continental and world estimates. Researchers may quarrel about the reasons and the methods for combining them and may argue for confining research to one or a few domains. This second purpose appears as the opposite of the first purpose, and this exemplifies the multipurpose nature of most samples. It appears also as the analog in another dimension for averaging over time.

3. *Averaging or aggregating over time* is a familiar concept in some ways but novel in others. Aggregating over the hours of days and over days of a week appears in many studies. The reference periods of many surveys (and censuses) cumulate production and consumption behaviors over the preceding year. Multiround surveys have been used to aggregate monthly births over yearly spans. But in many other surveys there may be an arbitrarily fixed reference point or span of time, though the collection time has to be more broadly spread. Most important, however, we shall argue for deliberate averaging over a longer span of time, e.g., over a year or even over several years (Section 6.1). These arguments for smoothing over seasonal and other temporal fluctuations are in addition to the need for cumulating over time in order to produce increased sample sizes and precisions for domains. But the relation of the two kinds of details—over space and over time—can be interwoven in balanced designs (6.1).

6.6C Strands of Change

We attempt to sort out the many possible changes that may complicate methods needed for combinations. Repeated simple random selections (SRS) from the "same urn" would be an ideal model, but too far from usual realities, and cumulations may be subject to important changes both in the population and in the sample. (For another perspective see 6.3 on sources of change.)

1. *Population changes* may be considerable, and these depend not only on the length of the period, but also greatly on the nature of the variables: economic variables and infectious diseases fluctuate much more rapidly than demographic variables. The composition and the size of the population will vary due to births and deaths, migration and other forms of mobility, net growth or attrition. Migration may be small across some national boundaries, but internal migration across internal boundaries can have much greater effect on regional and local statistics. Changes in internal boundaries can also occur more frequently and can complicate cumulations of data for cities and metropolitan areas, for example, and even more for organizations.

2. *Sample differences* must be distinguished from changes in the population itself. Differences between samples involve not only their relative sizes, but also the designs of the samples. In the case of similar designs for several samples, the relative precisions depend mostly on the sizes of the samples. But *similar* and *size* need proper technical definitions; for complex samples *design effects* and *effective sizes* need to be considered also (7.1). Biases due to nonresponses may be better considered under changes either in the populations or in the measurements.

3. *Changes in measurement* may be either planned or unplanned (6.3). They are especially troublesome for different sources of data; they are called "laboratory effects" in biological and chemical literature and are discussed at length in meta-analysis (6.6A4). In different situations or times even the same questions and methods may elicit different answers because of differences or shifts in meanings of words, concepts, responses, and even the kinds of nonresponses.

6.6D Types of Cumulations and Combinations

The following outline tries to list the distinct situations that seem most common or feasible, each with an example or two.

1. *Same population, same time, same methods.* Interpenetrating (replicated) samples are examples: A large sample is divided into k replications, each of which covers the entire population [Kish 1965, 4.4]. The method proposes to measure all variable (nonsystematic) errors with the variance between the k replications. The replications represent selections from the same population, distinguished only by sampling variation.

2. *Same population, same time, different methods.* One example can be a large survey sample divided between two (or more) organizations for field work. The results of the distinct surveys may be cumulated as sample cases, or the separately computed statistics can be combined later. Another

example would be combining data from the election polls conducted by several organizations about voting in the same election. The combined data would have details about domains (age, sex, occupation as well as geographic details), for which the sizes of the separate polls are inadequate. The separate surveys and polls should be examined for differences, but all those not obviously deficient can be combined when none are clearly superior (6.6B).

3. *Same population, different collection time.* The yearly cumulations of weekly samples of 1000 households of the National Health Survey [NCHS 1958] serve again as a simple example. Even here the stabilities of the population and of methods can be questioned; but those questions become more critical with greater differences in methods and with greater gaps in time. With different collection times we must distinguish two types of reference periods: the same *absolute* time of reference and the same *relative* time of references. The "number of children born last calendar year" (e.g., in 1985) would exemplify the same absolute time and could be cumulated weekly (e.g., over 1986). The "number of children born last month" collected each month exemplifies the same relative time; it is cumulated from "multiround" surveys in demographic surveys in some developing countries to reduce memory biases in reporting births. This involves aggregating and averaging over the time dimension.

This brings to mind the "age–period–cohort" connection that may also be illustrated with cumulation of births to mothers. A cross-section sample can obtain births during the past year from mothers of all ages. Yearly surveys can obtain past year's data for each age of mother, so that for each age of mother the cumulated data would come from different survey years. Instead, however, the cumulation can be by cohorts of mothers, so that a survey from year $x (x = 0, 1, 2, \ldots,)$ supplies data for that year x for age $a + x$. One year's survey can also supply data for several periods if memory can be trusted for retrospective data and if mortality is not a great problem.

4. *Different populations.* Here all the strands of change (6.3A, 6.6C) may operate. The types of cumulations noted earlier may be viewed as simpler situations, where the populations for the combinations could be considered as the same, similar, or stable. On the contrary, for combining samples from different populations, the differences between periods and/or between methods must also be considered. Among many possibilities we distinguish combinations over space and over time (6.6B). Most problems have been touched on elsewhere, but we merely add briefly combinations from *multiframe* coverage of the same population [Cochran 1964; Kish 1965a, 11.2D; Hartley 1974]. An example of *dual frame* coverage of households in the United States would be an area sample (A) plus a telephone listing (B). The entire sample covers $AB + A\bar{B} + \bar{B}A$ households but not the $\bar{A}\bar{B}$ households, which are missed by both the household and telephone frames.

6.6E Weights for Cumulating and Combining

This subsection differs from others, all of which are devoted to basic, often found problems and methods. But selecting weights for cumulating arises only rarely and recently, and this technical topic may be skipped until needed. It belongs to this chapter and I cannot find other treatments, as evidenced by the lack of references. We shall note several alternative methods for weighting, and the choice among them should depend on several criteria:

1. appropriate populations of inference;
2. validity and bias reduction;
3. simplicity and robustness;
4. efficiency and precision, including "measurability," that is, proper measures of sampling error (7.1E).

This would be my order of overall preference, but any of them could predominate in specific situations. As usual, there are no criteria for clear choice among criteria. Any situation would require examinations of the purposes for the pooling and of the likely strands of change and differences among samples (6.6C).

The alternative methods for weighting can be applied to (1) combining means with relative sizes p_g into $\bar{y} = \Sigma_g p_g \bar{y}_g$ with $\Sigma_g p_g = 1$ or to (2) cumulating individual cases with individual weights w_i into $\bar{y} = \Sigma_i w_i y_i / \Sigma_i w_i$. The sample means are $\bar{y}_g = \Sigma_j w_{gj} y_{jg} / \Sigma_j w_{gj}$, where $(g = 1, 2, \ldots, G)$ for samples and $(j = 1, 2, \ldots, n_g)$ for individuals, and $\Sigma_g n_g = n$. Within self-weighting samples the weights are $w_{gj} = 1$, and $\Sigma_j w_{gj} = n_g$ and $\bar{y}_g = \Sigma_j y_{gj} / n_g$. For cumulating individual weights $w_i = p_g w_{gj}$.

The relative weights p_g are useful for combining summary statistics like means, rates, proportions, and totals. They also facilitate choices among alternative methods in light of the preceding criteria. However, for cumulating individual cases into a cumulated sample of $n = \Sigma n_g$ cases the individual weights w_i offer more flexibility, especially for multivariate analysis. For complex statistics the individual y_i may represent vectors or functions of one or several variables, with moments such as x_i^a, $x_i^a z_i^b$ that need the weights w_i [Kish 1965a, 2.8C; Kish and Frankel 1974]. I doubt that "completely specified models" exist generally to eliminate the need for those weights (1.8).

Now we look at alternative methods for combining and cumulating. The discussion is entirely user oriented, and theoretical development would be helpful. In some situations the choice between methods may not be clear, and then one may compute two or more alternatives, compare them, perhaps even average them. There are many possibilities, and that is another reason we must neglect here formulas for computing variances for the pooled samples. The effects of overlaps between samples on variances are noted in 6.6F.

1. *Simple merging* of cases, with $w_i = 1$ for each case. Imagine several simple random samples n_g selected from the same population list. The cumulated mean should be simply the mean over the cumulated cases, $\bar{y} = \Sigma_g \Sigma_j y_{gj}/\Sigma_g n_g = \Sigma_g n_g \bar{y}_g/n$, and this makes $p_g = n_g/n$ for relative weights to combine the sample means $\bar{y}_g$. The cumulated mean can be regarded as selected with the combined fraction $f = n/N = \Sigma f_g = \Sigma n_g/N$. Here we assumed the samples were selected without replacement, hence without replications between the samples. If replications were allowed it may still be better to use this simple addition rather than to search and adjust for replications.

Whereas the simplicity of srs selections is not usually attained, it may be approximated. Several equal probability selections of elements from the same population may also be combined with $w_i = 1$ for each of the combined cases; for example, proportionate stratified element samples, even if the modes of stratification differ among the combined samples. However, if the samples have very different "design effects," efficiency (fourth criterion) may be better served by introducing corresponding differences into the weights, as in the following alternative 3.

Even for complex samples, if both the population and methods can be assumed to remain stable, simple cumulation may still seem reasonable. For k interpenetrating samples each selected with rates (probabilities) f/k, for the cumulated sample the cases can be assumed to be selected with rates kf/k. For combining 52 weekly samples selected with rates f, the NCHS [1958] can use simply $52f$ as the overall sampling rate, disregarding small changes in population size. Even differential probabilities within the surveys can be accommodated.

When, however, the populations or methods differ drastically between samples, other methods are needed for combining or cumulating samples.

2. *Population weights* can be assigned in accord with assumed population sizes $p'_g = N_g/\Sigma N_g$, instead of the sample sizes $p_g = n_g/\Sigma n_g$ earlier (1). This approach may be justified when population differences seem more important than sampling precision (e.g., for large samples and censuses). Population size N_g is only one measure of relative importance and other relative weights of importance may be considered instead. The accuracy or reliability of measurements may also be used as the basis for assigning relative weights to samples.

The most obvious weights would be equal $p_g = 1/G$ for each of G samples, and weights proportional to population sizes ($p_g \propto N_g$) seem to be only one reasonable departure. Another possible departure would assign a decay factor σ for each receding period, so that weights would be proportional to 1, $(1 - \sigma)$, $(1 - \sigma)^2$, etc., as sample periods recede. Then $\sigma = 0$ represents equal weights $1/G$; and $\sigma = 1$ represents giving full weight 1 only to the latest and zero weight to all others (a common practice). If each extreme ($p_q = 1/G$ and $p_g = 1$) may seem reasonable in different situations, a compromise with $0 < \sigma < 1$ should be sometimes better than both.

These considerations concern models for sources of variations over the span of the combinations. They concern the first two criteria—appropriate populations of inference and validity. On the contrary, the chief considerations for the following methods are efficiency and precision, along with simplicity and robustness.

3. *Effective sample sizes* $p_g \propto n'_g$ can yield greater precision than simply accepting $p_g \propto n_g$ when there are considerable differences in the efficiencies per element between the separate samples n_g (7.1). The effective sizes $p_g \propto n''_g = n_g/\text{deft}_g^2$ may be a first suggestion where the deft_g^2 (design effects) differ greatly because of great differences in clustering. However, this alternative fails the third criterion of simplicity and robustness because deft_g^2 differ greatly for diverse variables, for diverse statistics, and for diverse subclasses; it is not simple to assign separate weights w_i for each case for each of these statistics, especially for multivariate statistics. Some compromise average value may give most of the gains, without attaining optimality. For multivariate and for subclass statistics the design effects tend to be reduced, hence also relative disparities between the deft_g^2, hence also between optimal values for the n''_g. We may be more willing to sacrifice optimal precision than validity.

One source, not uncommon, of inefficiency comes from variations in weights of w_{gj} among cases within the same sample n_g. If these variations are haphazard (rather than optimal), they tend to increase variances in comparison with *epsem* (equal probability) selection. Then we may use $n''_g = n_g/[1 + CV^2(w_g)] = n_g/[\Sigma_j w_{gj}^2/(\Sigma_j w_{gj})^2]$ as the "effective size," where $CV^2(w_g)$ denotes the squared coefficient of variation of element weights w_{gj} within the sample n_g. If the range of relative weights is less than 1:1.5, the adjustments are trivial (e.g., for nonresponses or poststratification). However, disproportionate allocations for domains (regions, ethnic groups) or frame problems may cause greater ranges of variation. If differences between samples of $CV^2(w_g)$ are greater, then weights $p_g \propto n''_g$ may differ greatly from $p_g \propto n_g$ and justify their use [Kish 1976; 1965a, 11.7].

4. *Weights* W_i *proportional to* $W_{gj} \propto 1/P_{gj}$, inversely proportional to the probabilities of section P_{gj}, seem like the most "natural" method for cumulating elements j from samples g. But we face practical problems: We need to know and use the selection probabilities *within each of the G selections for each element that appears in the combined sample.*

Simple merging of unweighted cases amounts to assuming final case weights of $w = 1/f = 1/\Sigma f_g = 1/\Sigma(1/w_g)$, with uniform sampling rates $f_g = 1/w_g$ within samples, from the same "superpopulation." These weights estimate population totals, and they estimate means when divided by N.

The weights could also be extended to unequal probabilities $w_i = 1/P_i = 1/\Sigma_g P_{gi} = 1/\Sigma(1/\Sigma_g W_{gi})$, provided we could learn and use the probabilities P_{gi} for each element i that appears in the combined sample within each of the selections g. However, secure knowledge about the reasons for the unequal P_{gi} is unlikely and can lead to (serious) biases. For the case gi from the gth sample how can one tell what weight it should have in the imaginary superpopulation? Knowing and using P_{gi} can be difficult in the general case, with changes in the population composition, in the sampling methods, or even in some measurements. For example, $P_{gi} = 0$ is possible in some samples, either because of sampling rules of inclusion or because of population changes (births, deaths, migrations, boundaries). If the missing domain can be clearly distinguished, its weights can be based on other samples (g') only, $W_d = 1/\Sigma_{g'} f_{g'd}$.

If selection rates f_g can be fixed for each sample together with adjustment factors C_{gd} for each domain d, then the weights $W_d = 1/\Sigma_g C_{gd} f_g$ can be computed; the adjustment factors can be assigned for differential nonresponse, for poststratification, for oversampling domains, etc. However, reliance on stable knowledge of the factors C_{gd} could lead to biases, especially because of transfers into and out of domains (migration, mobility, birth, deaths, etc.).

Without fairly secure knowledge about the P_{gi}, it may be the better part of valor to yield some of the precision that this method could yield and rely instead on the more robust methods 2 or 3.

6.6F Gains from Combining Overlapping Samples

When combining samples that are independent, the precisions increase as variances decrease proportionately with the sizes of the samples in a straightforward manner. However, for overlapping samples the situation becomes more complicated and the gains from combinations tend to be less in proportion to correlations from the amount and the kind of overlaps (6.2B). The variance for the combined mean of two equal samples would be $(S^2 + S^2)/4 = S^2/2$ with nonoverlap; $(S^2 + S^2 + 2S^2PR)/4 = (1 + PR)S^2/2$ with partial (P) overlap and $(1 + R)S^2/2$ with complete overlap ($P = 1$).

For means based on J overlapping samples the general expression would be

$$\text{Var}(\Sigma \bar{y}_j/J) = (\Sigma S_j^2 + \Sigma S_j S_k P_{jk} R_{jk})/J^2, \qquad (j \neq k).$$

If we assume uniform S_j^2 over all samples this becomes $(S_j^2/J)\left(1 + \sum_{j \neq k} P_{jk} R_{jk}/J\right)$, and the last factor is a minimal 1 for $P = 0$ nonoverlaps only. There are $J(J - 1)$ terms in the covariances and the combined variance becomes $(S_j^2/J)[1 + (J - 1)R^*]$, where R^* represents an average correlation term for the $P_{jk} R_{jk}$ among all the J pairs. This would be a complex relationship where R_{jk} probably fluctuates between periods but also decays over longer spans. Furthermore, for partial overlaps the P_{jk} may also be large between neighboring spans but may tend to vanish for larger spans. Thus the covariances may be much more important for closer than for distant pairs of periods. However, in case of complete or large overlaps this term can be large for large values of R^*. Thus the mean of J samples may have variances anywhere from S_j^2/J for nonoverlaps or for $R^* = 0$ to approaching merely S_j^2 for complete overlaps and very high values of R^*.

CHAPTER 7

Several Distinct Problems of Design

I do not choose to speak more clearly than I think. Niels Bohr.

You don't need the masterpiece to get the idea. Pablo Picasso, Primitivism in Modern Art.

The best is the enemy of the good, Voltaire.

Anything worth doing is worth doing badly. Lord Chesterfield quoted by G K Chesterton.

These six sections may best be viewed as appendixes to the main body of the book. Each section concerns a distinct topic with little connection between them. Rather they connect with various places in the six chapters, where references to these appended sections abound. These technical topics arise often and their treatments were postponed in order to prevent their interrupting the main flow of discourse.

One common characteristic of all these six subjects is their special and *technical* nature. Because of this they do not receive adequate or any treatment in ordinary textbooks, and I cannot assume the readers of this book to be familiar with or even aware of them.

Nevertheless, I judge them to be important problems of *design*, and that constitutes their second common characteristic. Leaving them out entirely would make the treatments and decisions of this book seem arbitrary, incomplete, curtailed. The choice of these topics may also seem arbitrary; others could be added, but I had to stop somewhere.

A third characteristic of these six topics is that a *nontechnical treatment* of each seems both possible and useful. The discussion of each topic is necessarily incomplete, with references to more technical and complete treatments—where I could find them. However, some topics still await adequate treatments. Yet none of these forms of incompleteness would justify my leaving the reader unaware of the existence of these important problems.

7.1 ANALYTICAL STATISTICS FROM COMPLEX SAMPLES

7.1.A Limits Around the Topic

The title refers to a twofold complexity of statistical estimates and of selection methods; hence the topic could be potentially vast and vague. For a concise treatment we introduce several limitations. These can be amplified with references to other treatments for those problems that are excluded from here. This section continues the discussion of Chapter 2.

1. *Complex Analytical Statistics.* Discussions in Chapter 2 on "Analytical Uses of Sample Surveys" concern estimates of means and of their comparisons, or differences ($\bar{y}_c - \bar{y}_b$). These cover most of the presentations of data in social research even today. Nevertheless we have seen a continuing expansion in social research of using more complex analytical statistics, which is greatly facilitated by the spread of electronic computing. That expansion easily outran the foundations in statistical distribution theory, especially those for complex analytical statistics. These refer here to statistics more complex than comparisons of means; regressions and multivariate analysis in general would be good examples. In terms described in Figure 2.2.1 they refer chiefly to column 3 of that Figure, but the discussions of Sections 2.6 and 7.1C can also enlighten our view and treatment of means and of their comparisons in columns 1 and 2.

2. *Cluster Sampling Versus Stratified Element Sampling.* Figure 2.2.1 also distinguishes the effects of stratified element sampling on row B from those of cluster sampling on row C. These two kinds of departures from independent (simple random) sampling usually have entirely different effects on the variances of statistics. Proportionate stratified selection of elements tends to *reduce* the variances of means, but only *slightly*. The design effect (deft) ratio is slightly less than 1 (2.5). Those reductions tend to disappear from comparisons of means and to assume that deft $\simeq$ 1 (negligible effects from proportionate stratification) for them is not misleading (cell B2). Similar conjectures of negligible effects for complex analytical statistics (cell B3) usually are also good approximations, and those conjectures have been supported both with theory and with empirical results.

The effects are usually quite different for cluster sampling: Clustering tends to *increase* the variance; the increases can vary a great deal, and *may be large*; and the increases persist (even if reduced) for subclasses and for comparisons, also for complex analytical statistics (cells C2 and C3 in Figure 2.2.1). With deft^2 denoting the ratios over simple random variances, we say that deft^2 is greater than 1, sometimes only slightly but often considerably, and higher for means than for their comparisons or for analytical statistics. Those increases and conjectures have been justified both theoretically and empirically (2.6, 7.2B.9).

3. ***Unequal Weights Versus Lack of Independence.*** The effects of complex selection methods (clustering and stratification) that serve as controls (restrictions) on the independence of sampling probabilities have been dealt with often in sampling literature (also in Chapter 2). The effects of weights introduced to compensate for unequal selection probabilities have been mostly neglected and are quite different from those due to stratification and to clustering.

One reason for that neglect is due to the prevalence in research of self-weighting samples, when the elements are selected with equal probabilities; and there is much to recommend them [Kish 1977]. There may be slight departures from equal probabilities due to (1) adjustments for nonresponse and noncoverage; or (2) other statistical adjustments for sampling variation; or (3) small inequalities in selection frames and procedures [Kish 1965a, 2.7, 11.7; Verma et al. 1980]. Examples of (3) are weighting for numbers of adults in households when either addresses or telephones are selected with equal probabilities [Kish 1965a, 11.3; Groves and Kahn 1979]. Such departures generally have, or should have, mild ratios to unity and consequently only mild effects on the variances. However, these "mild" effects are not entirely negligible, perhaps amounting to 5 to 30 percent increases in the variances. Furthermore, these variance increases tend to be "inherited" in the subclasses, in comparisons and in analytical statistics: They do not decrease (7.4).

Quite different from those "mild" departures are drastic and deliberate departures from equal probabilities for disproportionate or "optimal" allocation of sample cases. Three distinct examples will help to illustrate these: (1) deliberate oversampling of a minority subclass to achieve equal sample sizes (e.g., oversampling by the factor 10 to achieve "equal" samples for 10 percent of blacks in the United States); (2) "optimal" allocation for rare but important elements (e.g., large stores or institutions) [Kish 1965a, 3.5]; (3) selection with PPS (probability proportional to size) for "units of variable sizes" (7.5). Departures in all these cases can and should be large, and the necessary weighting procedures should be deliberate and specific to the situation (7.4).

Faced with the possible increases in variances due to weighting of sample cases, would it be better to forego the increases in both variances and in complexity of analyses and to compute unweighted estimates? We need not repeat here the reasons for weighting up to population values, presented elsewhere (2.1, 2.7, 7.4).

4. ***Descriptive (Point) Estimates Versus Inferential Statements.*** This section focuses on the effects induced by lack of independence on inferential (probability) statements, especially on the increases due to clustering in the sampling errors of analytical statistics. Thus we follow the topics discussed in Section 2.2, that descriptive statistics, "point estimates," or "*first-order*" *estimates are not affected, or only slightly affected by the lack of independence* in the selection (by differences in joint probabilities) between population elements. This signifies that for samples of moderate or large sizes the descriptive statistics (properly weighted, if necessary) are "good" and "consistent" estimates of corresponding population values. This statement holds for simple sample statistics like the mean $\bar{y}$ and the element variance s^2; it holds also for complex analytical descriptive statistics, such as regression coefficients.

7.1B Alternatives for Sampling Errors of Complex Samples

The following distinct alternatives for computing sampling errors are not equally appropriate to all specific situations. However, two or three of them may all seem reasonable in any situation, and a clear choice between them may not be obvious. One of them should not be accepted lightly as the obvious choice, without considering other alternatives. Sometimes, alas, a better choice emerges only with hindsight, after the results are in. None of these alternatives should be dismissed out of hand, because they all have had their uses. But frankly, my chief interest lies in the last alternative, where proper computations are combined with approximations to obtain reasonable sampling errors for analytical statistics from clustered samples (7.1C).

1. ***Omit Computing and Presenting Sampling Errors.*** This practice, although in decline, is common even today. It is seldom defended theoretically, but sometimes along the lines I tried to present in 1.8.3. But for statistics based on censuses and on large samples connected with censuses (5.3), sampling errors may be less important than other kinds of errors and biases.

2. ***Use SRS Estimates for Sampling Errors, Disregarding the Complexity of Actual Selection Methods.*** These practices may well have increased with the availability of "canned" programs for analytical statistics, composed without advice from sampling specialists. In most programs the standard errors and tests of significance for those programs are computed with assumptions of independent selections, disregarding design effects. They may be good approximations for proportionate stratified element selections, but they can be gross underestimates for cluster sampling [7.1C].

3. ***Judgemental Inference to Population; Model Dependence.*** These may be based on naive disregard of design effects or on mathemically sophisticated attempts to hurdle obstacles. They may be represented by posing "internal validity" against "external validity" (3.5) and by approaches discussed in 2.1, 2.7, and 1.8.

4. ***Restrain Analysis; Wait for Distribution Theory.*** For means, subclass means, and their comparisons, statistical sampling theory provides adequate formulas for variances from complex samples; and a great deal of research is being and can be presented with ingenious multivariate utilization of those simple statistics. However, more complex analyses (regressions, log linear models, etc.) can have great analytical advantages; and to refrain from using them on complex samples (because these lack proper sampling errors) seems too great a sacrifice of knowledge contained in data. On the other hand, it would be futile to wait for an early development of statistical distribution formulas of errors for all those complex analytical statistics [Kish 1984].

5. ***Select a Simple Random Sample or a Reasonable Approximation.*** An srs may seem feasible when a good list of "all" population elements is available or feasible *and* location costs are not forbidding [Kish 1965a, 5.1]. For example, telephone surveys are frequently selected with telephone sampling, using "random digit dialing" in some countries [Groves and Kahn 1979]; these need some skill, care, and probably some toleration of imperfections, such as noncoverage, blanks, and unequal numbers of elements identified with the telephones. Also mail interviews from some literate and cooperative populations with good lists of adresses, such as the population registers in northern Europe, may be fairly successful. But even in these situations some of us would prefer to select an approximation to srs (such as a systematic or proportionate stratified selection), because it is easier as well as less variable; but these reductions of variances will probably be negligible for analytical statistics.

6. ***Select Simple Replicated (Interpenetrating) Samples.*** These methods would circumvent the obstacles raised by complex analytical statistics by returning to the basic statistical concept of simple, independent replications. That basic concept calls for *k independent* selections, but each selection

can be as complex as required by the situation. Then the variance of any statistics $\bar{b} = \Sigma b_j/k$, averaged over the k samples, may be computed as $\text{var}(\bar{b}) = \Sigma(b_j - \bar{b})^2/k(k - 1)$. These variances may also be computed as "jackknife" estimates.

The number of independent "interpenetrating" samples of $k = 4$ was advocated by Mahalanobis in 1948 and Lahiri. Clear description of the method for $k = 10$ replicates is given by Deming [1960, chs. 6–10], and Jones, in 1956, presented reasons and rules for 25 to 50 replicates. But in practice it is unsatisfactory for numerical reasons: if k is small (4) the computed variance lacks adequate precision; but large k (50) sacrifices too much of the control desired for good design (7.1E) [Kish 1965a, 4.4, with further references].

These obstacles seem even greater for designs of comparisons, concerned with "internal replications" of pairs of sites (3.1B, 3.1D). Lack of controls imposed by independent selections conflicts with the needs of controlled designs for "falsifiability" (7.6).

7. ***Use Repeated Replications, Jackknife, Bootstrap, or Resampling Methods.*** These names refer to a family of related methods for computing sampling errors when the number of independent replications is too small for useful estimates. Each method *repeats* the use of individual replications, thus *resampling* from the same selected units. Suppose we have only two selections from each of H strata, and such paired selections are common: They allow for valid estimates of variances with the minimal replication of 2 within each of the H strata. Thus $2H$ units are controlled by H strata, and H is often large.

For a simple example, however, we use only $H = 3$ and denote the three pairs of independent selections as $|Aa|Bb|Cc|$. Then the two independent halves ABC and abc would yield a simple replicated estimate of the variance of a statistic $\bar{b} = \Sigma b_j/6$ that is based on both halves, hence on the entire sample of six units. But two replications would yield a uselessly imprecise estimate of the variance, with only one "degree of freedom."

But with *repeated replications* we may use ABC, aBC, AbC, and ABc to compare with the overall mean (also abc, Abc, aBc, and abC, but these are mostly redundant). For two selections per stratum there are 2^{H-1} such repeated replications; for $H = 3$ there are 4, but for large H they would be too many. We may sample from that large number 2^{H-1}, and *balanced repeated replications* (BRR) permits this to be done efficiently.

For *jackknife repeated replications* (JRR) from the same three pairs one may use the b_j statistics based on the three combinations $AABbCc$, $AaBBCc$, and $AaBbCC$ and compare them with the overall statistics $\bar{b}$. From H pairs there would be H combinations with JRR.

Details, justifications, and comparisons are available for BRR, JRR, and linearization (Taylor, delta) methods [Kish and Frankel 1974]. All three methods can be used to compute roughly similar estimates of sampling errors, and none are clearly superior. More recently, "bootstrap" methods of resampling have been added [Efron 1982], but not yet developed for complex selection methods. For recent expositions see Rust [1984, 1986] and Rao [1986].

7.1C Approximations, Conjectures, and Analogies

Differentiate again the distinct problems B3 and C3 of Figure 2.2.1: These two differ greatly, hence class B3 for stratified element sampling is marked "conjectured," whereas C3 for clustered samples is "difficult." Let us discuss briefly the easier "conjectures" for proportionate stratified element sampling (pres). The conjectures stand on both theoretical and empirical legs and involve four steps. First, ample empirical data exist showing that only modest reductions of the variances can be obtained from pres, because most of the variance remains within strata even after stratifying the elements; hence $S_w^2/S^2 = \text{deft}^2 < 1$, but only by small percentages. Second, for crossclasses even those small gains are reduced in proportion to the relative sizes $\bar{M}_c < 1$ of crossclasses, and thus values of deft_c^2 tend toward 1. Third, for differences of crossclasses, the gains (reductions) in the variances tend to vanish altogether and $\text{deft}^2(\bar{y}_c - \bar{y}_b)$ approaches 1: The variance of $(\bar{y}_c - \bar{y}_b)$ approaches that for pairs of srs samples. These three steps (repeated from 2.5) combine theoretical with empirical results and lead to the fourth conjecture.

Fourth, it seems reasonable to conjecture that for complex analytical statistics the effects of proportionate element stratification will be negligible and that probability statements will be close to classical srs limits. The clearest (but not the only) evidence for these conjectures comes from eight diverse surveys in Israel on savings, attitudes, hospitalization, perception: the (iterated) ratios of pres/srs values of chi squares came to 1.0 or 1.00 in all eight surveys [Kish and Frankel 1974; Nathan 1972, 1973]. Differences between means measure relationships, and their variances approach srs. It is reasonable to conjecture that srs approximations will hold for other analytical statistics. It is a useful conjecture because appropriate computations of sampling errors for analytical statistics will be difficult for the foreseeable future. A cautious statistician faced with a large and important set of data could compute values of deft^2 for means of the sample and of crossclasses. If these show large departures from srs, then he may want to try some adaptation of the computing methods of repeated replications.

Effects of clustering on sampling errors of complex analytical statistics (C3 in Figure 2.2.1) pose problems that are more difficult, and more important, for several reasons. First, clustered and multistage samples are now common sources for complex statistical analyses. Second, the complexities of both the analyses and of the designs have many aspects, too many and too complex for mathematicians to develop distinct and useful distribution theories. Third, the design effects due to clustering are often both considerable and persistent, and ignoring them leads to serious overconfidence in sample results. Fourth, considerable design effects have been and are now reported widely for sampling errors of diverse analytical statistics.

Table 7.1.1 presents examples from three different sets of calculations, and we may note briefly several aspects, also noted elsewhere. First, design effects are considerably greater than 1, the srs value. Values of deft = 1.4 for standard errors (or deff = 2 for variances), for example, if ignored, would lead to gross underestimates of the real probabilities of erroneous inferences, because these are increased from $P = .05$ to about .15 and from $P = .01$ to about .07. Second, deft values within data sets are related and deft for complex coefficients seem to be always less than deft values for means: These latter can be computed to serve as reasonably conjectured upper limits. Third, the relations among the diverse coefficients (and other statistics) are not easily conjectured or predictable. For example, we wrongly guessed that partial correlation coefficients (because they are more complex, interactive) would have lower defts than simple correlation or regression coefficients; but they have not in Table 7.1.1. Results like these have been made possible by the emergence since about 1970 of repeated replications. They are being reinforced with similar results [Landis et al. 1982].

In many actual situations computing sampling errors for all (or even most) of the statistics presented just does not seem practical. Moreover, alternatives 5 or 6 above, selecting either an srs or a simple replication, are impractical. Ignoring or avoiding the problem with one of alternatives 1–4 is often practiced, but I cannot recommend any of them. Alternative 7 with methods of repeated replication or linearization is available for most situations but may be too difficult or expensive for some studies. In other studies sampling errors may be computed for some, but not for all the myriad statistics presented in their reports.

Thus approximations with conjectures seem inescapable; indeed they are widely practiced in connection with empirical results. They also appear in methodological investigations of the effects of complex selections. Alas, explicit general expositions and theoretical justifications are still needed. However, I shall describe some methods used for approximations, after a brief catalogue of the sources for analogies. We begin with relatively simple and safe conjectures and proceed to complex and bold methods as needed.

1. Simple situations of *small and rare clustering* may permit assumption of

TABLE 7.1.1. Deft = $\sqrt{\text{Deff}}$ for Standard Errors of Five Types of Statistics from Three Complex Samples[a]

Sample Set	A	B	C
Means	1.11	1.80	1.44
Simple correlation coefficients	1.10	1.26	1.36
Regression coefficients	1.02	1.30	1.11
Partial correlation coefficients	1.04	1.40	1.36
Multiple correlation coefficients	NA	1.46	1.89

[a]From Kish and Frankel 1974.

only negligible increase of deft over the srs value of 1. For example, suppose that dwellings are selected individually and with epsem; then for fertility surveys of mothers, the number of potential mothers in dwellings is seldom more than one, almost never more than two. This is only a little less true for surveys of adults of either one sex in developed countries with small nuclear families. Even in samples of small area segments, the members of a widely distributed and rare *crossclass* may appear mostly as individuals.

2. Relatively safe conjectures may be made from *the same statistics from repeated studies* with similar sample sizes, designs, and variables. Constancy for relations like homogeneity provides a reasonable basis for conjecture for the same statistics over the series of similar repeated studies.

3. However, the analogies become weaker, hence models must be stronger, when we must make conjectures not from the same statistics but from "*similar*" *statistics from the same survey*. Finding and defining "similar" statistics may seem difficult, but not entirely impossible [Verma et al. 1980, Kish et al. 1976]. For example, it may seem reasonable to impute values of deft for standard errors of regression coefficients from other similar regressions from the same study. These other regressions may be chosen to be shorter and simpler to compute, in order to save effort on the larger ones.

4. We proceed to bolder analogies if we must. Values of deft($\bar{y}$) for *means*, which are most easily computed, may be used as *upper bounds for more complex statistics* from the same studies, as discussed later in more detail.

5. *Borrowing sampling errors from "similar" studies* is fraught with risks and needs the attention of good statisticians to judge what is "similar" and to advise on methods for translation from the source of computations to the destination of needed sampling errors. Nevertheless this is common practice and probably better than simply assuming srs, when alternatives 2, 3, and 4 are not available. Other studies and methodological investigations are the sources for models and relationships for the needed conjectures.

7.1D Methods and Models for Sampling Errors

Computations of sampling errors for surveys are generally multipurpose in two ways: First, they concern many statistics for many variables; second, they can serve several different needs. Computing sampling errors involves more than variances and standard errors. Design effects (deft^2), ratios of homogeneity (roh) and coefficients of variation (especially $cv(x)$ in 7.1E) are useful and used; from these, averages and other functions can be computed also, and they often are.

Design effects, the ratios deft^2 = actual var/srs var, have several important uses. (1) They may be averaged for greater stability, when the computed variances are subject to great variation, because they are based on few primary units, or "degrees of freedom" (7.1E). (2) They can and should be used to check for gross errors in variance computations. Gross errors are the most common and easiest to spot for deviations from the base of 1. (3) Their main purpose is in models and conjectures for other statistics from the same survey. But for this purpose the function·roh = $\text{deft}^2/(\bar{b} - 1)$ is preferable, especially for crossclasses. (4) They may be "borrowed" to serve in conjectures about sampling errors for other surveys. (5) They may be used for designing other surveys.

Some simple rules (a–f), relationships, and models follow for computing sampling error functions and for their broad, multipurpose uses.

a. *Compute ste($\bar{y}$) and deft($\bar{y}$) for many means.* Variances for means (and proportions) are relatively easy to compute, and with today's programs they can be readily computed simultaneously for many variables for the overall means ($\bar{y}$). At the same time srs variances can be computed, and from the ratio deft^2 = actual variance/srs variance, values of deft should be obtained for the full diversity of variables covered by the study. Although they will exhibit a much smaller range than standard errors, the deft values often still differ a great deal for different variables. Three kinds of situations may ensue in cluster sampling: (1) All the deft may be near and mostly higher than 1.00, pointing to low, perhaps negligible, design effects; (2) a great deal of variation may exist both above and below 1.00, pointing to too much variation for the estimates of the variances due to too few replicates, "degrees of freedom" (see item f); (3) a fair amount of variation may be found for deft values above 1.00, perhaps up to 2, 5, or even beyond, and this is the situation that most needs to be treated.

These values of deft($\bar{y}$) for means of different variables from the entire sample can each be rich sources of conjecture for other statistics. The following procedure depends on models of relations between diverse statistics for the same variable. However, there often are great differences between variables, and it is difficult to model good conjectures between variables.

b. *Deft.* For measures of sampling errors, researchers most often want to use something like standard errors. These are known best in the forms $\text{ste}_0(\bar{y}) = s/\sqrt{n}$, or $\text{ste}_0(p) = \sqrt{pq/n}$ for srs. These and similar standard errors $\text{ste}_0(g)$ are also available these days as standard outputs of computing programs for many complex statistics (g). For example, the outputs for multivariate linear regressions yield not only the regression coefficients g_k but also standard error $\text{ste}_0(g_k)$ for those coefficients. However, these are all srs estimates that may seriously underestimate the actual errors of those g_k statistics. We need the factors $\text{deft}(g) > 1$ that will bring the standard errors closer to their proper value: $\text{ste}(g_k) = \text{deft}(g) \times \text{ste}_0(g_k)$. The utility of deft for conjectures and analogies is due to its relative stability by removing three sources of confusion ("nuisance parameters"): units of measurement, the spread of the frequency distribution (σ^2), and the overall sample size (n). Thus the defts are more "portable" than standard errors for comparisons over statistics and across studies.

The best source for "borrowing" such deft(g) values would be "similar" statistics for similar variables. But if these are not available we may also make use of $\text{deft}(\bar{y})$ values for means, with this relationship:

$$1 < \text{deft}(g) < \text{deft}(\bar{y}) \tag{7.1.1}$$

There is a fair amount of empirical evidence and methodological justification for this statement: The variance of analytical statistics (e.g. regression coefficients) in cluster samples, though greater than 1, tends to be less than the $\text{deft}(\bar{y})$ for means from the same database. These relations yield sufficiently narrow limits when the values of $\text{var}(\bar{y})$ are not far from 1.00 and not too diverse.

c. *Deft² and roh.* For means of crossclasses the effects of clustering should decrease as the size of crossclasses becomes small, and deft^2 should approach 1.0 as the average cluster sizes approach 1. If the value of $\text{deft}^2(\bar{y})$ for the entire mean is already close to 1, there is little room or need for conjectures. However, if $\text{deft}^2(\bar{y})$ is substantially above 1, it is comforting to know that a wealth of evidence bears out the simple conjecture of almost linear decreases of $\text{deft}^2(\bar{y}_c)$ for crossclass means ($\bar{y}_c$). This follows from $\text{deft}^2 = [1 + \text{roh}(\bar{b} - 1)]$ and from the conjecture that roh remains relatively constant from roh_t for the total sample to roh_c for the crossclass (2.6). Then $\text{deft}(\bar{y}_t) = [1 + \text{roh}_t(\bar{b}_t - 1)]$, and $\text{deft}(\bar{y}_c) = [1 + \text{roh}_c(\bar{b}_c - 1)] = [1 + \text{roh}_t(\bar{b}_c - 1)]$. The last step assumes $\text{roh}_c = \text{roh}_t$; but actually roh_c tends to be slightly higher than roh, even more than slightly for subclasses that tend to be clustered and segregated, such as socioeconomic subclasses [Kish, Groves, and Krotki 1976; Verma et al. 1980]. Hence these approximations must be used with some caution, and only when the cluster sizes are fairly even (Figure 7.1.1a).

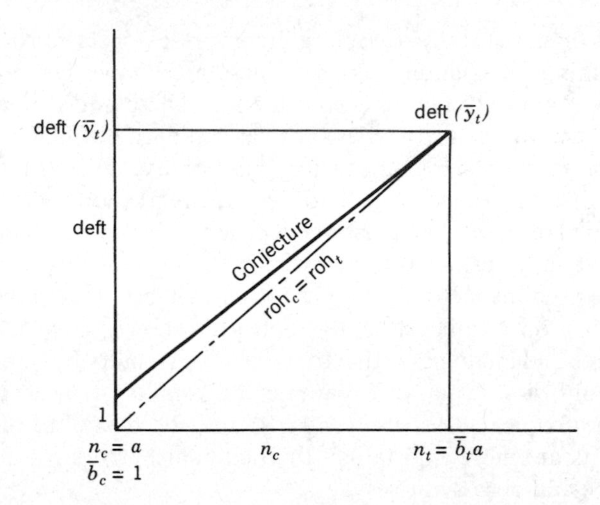

Figure 7.1.1a. Roh values are more "portable" from one crossclass to another than deft² values, which are functions of the size $\bar{b}_c$ of the crossclasses in the clusters, since $\text{deft}_c^2 = [1 + \text{roh}_c(\bar{b} - 1)]$. But the values of roh_c seem to remain almost constant from roh_t on down to small crossclasses. Therefore deft_c^2 for crossclasses declines almost to 1.0 as the crossclass sizes decrease from the total $n_t = an_c$ to $n_c = a$, that is, the cluster sizes decline from $\bar{b}_t = n_t/a$ to $\bar{b} = 1$. But between variables the values of roh can vary a great deal, and variations from 0.001 to 0.200 are common.

Nevertheless, for inference to crossclasses these approximations are much better, when deft is not negligible, than assuming either than deft is constant at deft = 1 (as in srs) or that $\text{deft}(\bar{y}_c) = \text{deft}(\bar{y}_t)$.

Thus for our approximations for crossclass means we may use two stages of conjecture: first ste to deft, then deft² to roh, then back in two stages of roh to deft² and deft to ste. (See Figure 7.1.1b.)

Roh values have greater relative ranges of variation than deft values: even a roh = .001 may be nonnegligible with $\bar{b} = 300$ (since $1 + .001[300 - 1] = 1.3$), but values as high as roh = .200 are not uncommon. But roh values are much more "portable" to crossclasses than deft values. They are also more "portable" across studies with different designs and even across populations for the same variables; and for "similar" variables with more caution.

Design effects for differences between crossclasses may decrease toward 1.00 due to covariances within sampling units, but in a manner too complex to discuss here (2.6).

d. *Effective n.* In the variance of the mean from a complex sample $\text{var}(\bar{y}) = \text{deft}^2(s^2/n)$, we viewed deft² as acting on the variance of the mean s^2/n. We also could regard deft as acting on the element standard deviation s

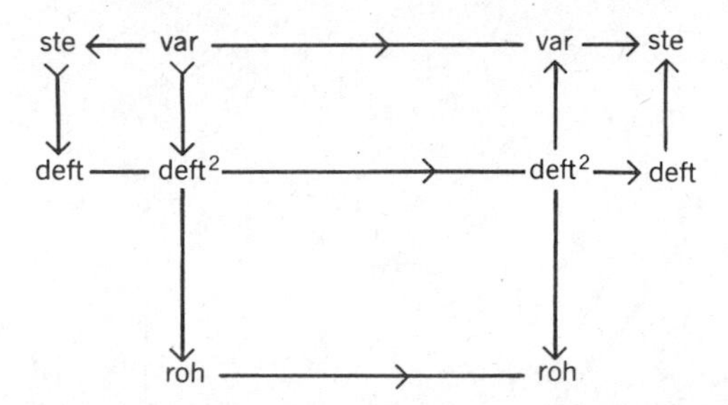

Figure 7.1.1b. Cluster sizes vary greatly both between samples and between crossclasses used in analyses of the samples. For the reasons in 7.1.1a, it is more reasonable to use the three-level inference from variances to deft² to roh and back again than a two-level inference through the deft². Variances and standard errors are even less portable, because they depend also on units of measurement. This justifies the need for seven steps from standard errors to standard errors through the rohs.

For differences $(\bar{y}_c - \bar{y}_b)$ of two crossclasses, the declines of deft² toward 1.0 are further accentuated by subtractions of usually positive covariances.

and $\text{var}(\bar{y}) = (\text{deft } s)^2/n$. But this approach becomes involved for complex analytical statistics. However, it may be more helpful to regard deft² as acting on the size of the sample n. Thus $n(\text{effective}) = n/\text{deft}^2$; that is $\text{deft}^2 > 1$ reduces the effective sample size proportionately. This has been used conveniently for tests of significance [Rao 1986].

e. *Effects of haphazard weights.* The effects of haphazard weights on the variances tend to remain undiminished in crossclasses, comparisons, and in analytical statistics. This differs from the design effects $\text{deft}^2 < 1$ for proportionate stratified element sampling which tend toward 1 from below, and the $\text{deft}^2 > 1$ for cluster sampling which tend toward 1 from above (Figure 7.1.2).

f. *Components of the variance.* Conjectures and approximations based on values of deft and roh "yield rough-and-ready," "quick-and-dirty" methods for dealing simultaneously with several complexities of survey samples. They permit computing for and handling of many variables and many statistics, and those are our foremost tasks.

However, this overall method hides the full complexities of multistage samples. For example, a three-stage sample with stratification at each stage may have six or more components. But all those components are seldom computed, especially for the many variables that most surveys need, and for several reasons. First, the work would be complex and consuming. But second and more important, subtracting several components yields values that are highly, even uselessly, variable. Third, it would be too difficult to present and to interpret many components for many variables.

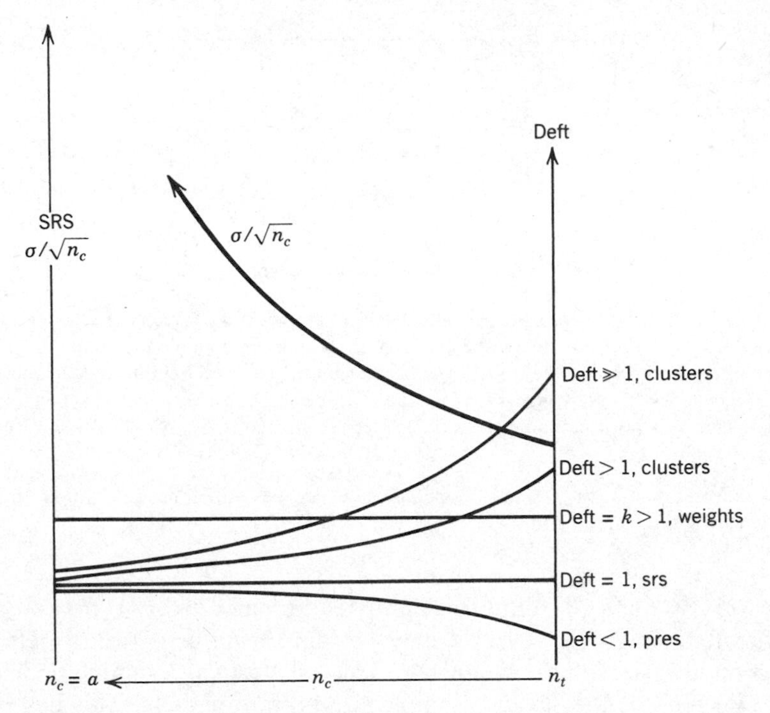

Figure 7.1.2. Convergences of deft to 1 for decreasing crossclass sizes.

When the crossclass size decreases from the total sample size n_t toward $n_c = a$ (the number of clusters), the average cluster size decreases from $\bar{b}_t = n_t/a$ toward 1. The simple random standard error $\sigma/\sqrt{n_c}$ increases by the factor $\sqrt{\bar{b}_c}$ where $\bar{b}_c = n_c/a$. But the design effect deft > 1, due to clustering, decreases toward 1, at the same time (Figure 7.1.1b), hence the complex error deft $\sigma/\sqrt{n}$ increases by less than $\sqrt{\bar{b}_c}$. For differences of crossclass means, the effects tend to be further reduced toward 1.

We note three other possible effects for decreasing crossclasses. The deft < 1 for proportionate stratified element samples (pres) generally increase toward 1. The effects due to "random" weighting, small, if any, tend to remain constant (k). The effects on design subclasses tend to remain constant, on the same level as for n_t on the average, but they may be different and difficult to generalize.

g. *Models.* A great variety of models, other than defts and rohs, may become available eventually. Now, however, we can point only to the use of the coefficients of variation, $\text{cv}(\bar{y}) = \text{ste}(\bar{y})/\bar{y}$. With the conjecture of constant values of cv, we can use $\text{ste}(\bar{y}_d) = \bar{y}_d[\text{ste}(\bar{y}_c/\bar{y}_c]$ to conjecture from the computed values for $\text{ste}(\bar{y}_c)$ to the desired $\text{ste}(\bar{y}_d)$. This model removes the units of measurement and depends on the means and standard errors (ultimately the standard deviations) being proportional between variables. It has been used with success for positive variables skewed toward higher values, but it may perform poorly for others [Hansen et al. 1953, Sec. B15].

Models are also needed for averaging and generalizing of sampling errors (standard errors, defts, rohs) computed with few replications (degrees of freedom) and hence subject to excessive variability. Expositions and descriptions are sparse, but some examples may help [USCB 1978; Gonzalez et al. 1975; Verma et al. 1980].

7.1E Measurability of Sampling Errors

"*Measurability* denotes designs which allow the computation from the sample itself, of valid estimates or approximations of its sampling variability" [Kish 1965, 1.6]. This basic idea permeates survey sampling and lies in the foundation of experimental design. R A Fisher wrote [1935], under "26. Validity of the Estimation of Error," about replication: "its main purpose, which there is no alternative method of achieving, is to supply an estimate of error by which the significance of these comparisons is to be judged.... The purpose of randomization is to guarantee the validity of the test of significance, this test being based on an estimate of error made possible by replication." He also wrote, under "20. Validity and Randomization," that "our estimate of the error of the average difference must be based upon the discrepancies between the differences actually observed ... to guarantee that such an estimate be a valid one...." Similar concepts underlie designs of samples and of experiments with four basic aspects.

1. The computation of error estimates must be based on variations found in the sample data themselves. Survey samplers do not assume srs selection for clustered samples, and $deft^2$ measures the discrepancy. Those who would deny this strong connection are heavily "model dependent" (1.8).
2. Randomized replicates are the bases for valid error estimates, and the computations of errors must reflect the units randomized in the design. The error estimates must reflect the clustering and stratification (blocking) used in the design. The "first-order" statistics can be computed from knowledge of the probabilities p_j of selection of each sample element. However, computations of the "second-order" measures of sampling variability are affected by the pairwise joint probabilities p_{ij} of the elements, and these depend on the complexities of design.
3. When valid estimates are difficult in practice, "good" approximations may be substituted, and to that extent the computations become "model dependent," as examples below show.
4. We discuss sampling variability and sampling errors, not only sampling variance, in order to include standard errors and mean square errors;

other aspects of sampling errors, such as design effects and coefficients of variation; confidence intervals and other probability intervals of uncertainty and tests of uncertainty; and variable, measurable errors of observation.

The *practical requirements for a measurable* design in survey sampling are not many.

1. The design must have two or more randomized replicates selected from each stratum in order to permit computation of variances from each stratum.
2. To have useful precision for sampling errors it is best to have about 30 or more "degrees of freedom." For example, for a design of two counties (areas) from each stratum in the primary selection stage, one would need $30 \times 2 = 60$ counties (areas) in the whole sample; each stratum yields one degree of freedom.
3. Identifying numbers of the strata and primary selection units must be available for all sample cases on the tape (or disc, deck, list). This obvious need has been *usually* neglected in practice, thus making it impossible to compute valid sampling errors.
4. The preceding identification (3) is sufficient for computing overall sampling variability from primary selections, often called "ultimate clusters" [Kalton 1979; Kish 1965a, 6.5; Hansen, Hurwitz, Madow 1953, 6.7]. This has great practical advantages in multistage sampling: For the stages after the primary stage the identifying numbers of strata and of sampling units may be ignored.
5. Those numbers for later stages would be necessary only for computing components of variances for later stages. But those components are seldom computed, both because they would be complicated and because they would be unstable. It would begin with an unstable overall variance from the first stage (because this is typically based on small number of primary selections), and it would become progressively less stable as variable components have to be subtracted for successive stages of selection. Estimates of the components would be useful for designing future samples, but not necessary for inferences for the sample results.
6. We had to abandon strict rules in order to have practical guides to measurable samples. First, we could not insist on unbiased estimates of the variance, but were satisfied to accept mean squared errors with tolerably small biases (7.2F). Second, our measurable designs will not satisfy all estimators, because for some estimators no design may yield

reasonable error estimates. We need a practical definition for designs that will yield measurability for most common estimates, especially linear estimators like means.

Some diverse examples should help clarify concepts and problems, perhaps with the help of a few definitions that follow immediately below.

1. Consider a sample of a single unit that is a cluster of elements—e.g., a single district (or county or school) to represent a state (or country). This has no measurability (beyond the single unit) because variations between persons (regardless of how many) fail to reflect variations between such units in the population.
2. A sample of two clusters from the population may be judged measurable, because an unbiased variance estimate (UVE) can be computed from the sample itself. But this estimate is so variable as to be almost useless, and I prefer that "valid estimates" in the definition of *measurable* also imply "useful" estimates, although this requirement leaves us without clear boundaries for *useful*, *valid*, and *measurable*.
3. What then about "interpenetrating" samples with $k = 4$ or $k = 10$ independent replications? Variances computed with 3 degrees of freedom are almost useless in most situations, I believe, and 9 df are marginal and need "pooling" of some kind [Kish 1965a, 4.4].
4. Many good samples contain only single primary selections from each stratum, and then they are usually paired into "collapsed strata" to yield the differences needed for variance computations. These methods yield variances that are judged to be only slightly and tolerably overestimates, as described in most textbooks under "collapsed strata" [Cochran 1977, 5A.16; Kish 1965a, 8.6B; Hansen, Hurwitz, and Madow 1953, 10.13].
5. Systematic sampling of primary selections presents a problem that in practice is also treated adequately with "collapsed strata" of neighboring selections. However, because the entire design is determined with the selection of the single starting random number, this is a probability sample that is theoretically not measurable. The variance computations depend on a model of randomness within neighboring strata [Cochran 1977, 8.10; Kish 1965a, 6.5C].
6. These remarks about measurability concern most of the commonly used statistics—such as totals, means, subclass means, and comparisons—in columns 1 and 2 of Figure 2.2.1. Furthermore, they can be extended to the complex statistics (coefficients in regressions) in column 3 by using methods of repeated replications (resampling, BRR,

JRR), even when no mathematical expressions for the variance are available. However there may be statistics for which even these robust and general methods fail to work and no valid or good approximations are yet available (7.1.B).

7. Measurable samples are often rendered nonmeasurable by failing to maintain proper records to identify the numbers of the strata and primary selections for the sample cases.
8. What about k "identical" replications of the same nonprobability (judgment, quota) sample? Although a variance could be computed around its own mean (expected $E(\bar{y})$) value, this has an unknown gap to the population mean $\bar{Y}$, hence I judge this model to be not measurable.

Justifications for the preceding statements are scattered in the literature of survey sampling, and the indexes refer specifically to collapsed strata, systematic sampling, degrees of freedom, and similar terms. Only a few basic ideas may be injected here, beginning with the familiar example of the mean from a simple random sample, where $E[(1 - f)s^2/n] = (1 - f)S^2/n = E[(\bar{y} - \bar{Y})^2]$. This exemplifies a general statement for the variance of some statistic $\bar{y}$: $E[\text{var}(\bar{y})] = \text{Var}(\bar{y}) = E[(\bar{y} - \bar{Y})^2]$. The first equation denotes designs where the computed sampling variance is an unbiased variance estimator of the mathematically derived expression for the population variance, based on population parameters (S^2). The second part of the equation denotes that this analytical variance expresses the expectation of the mean squared errors of the statistic $\bar{y}$ from the population value $\bar{Y}$. This MSE for the sampling distribution of $\bar{y}$ is needed so that $\sqrt{\text{var}(\bar{y})} = \text{ste}(\bar{y})$ may be used with specified confidence in statements like $\bar{y} \pm t_p\text{ste}(\bar{y})$.

When $E(\bar{y}) = \bar{Y}$, then $\bar{y}$ is an unbiased estimator of $\bar{Y}$ and $E[(\bar{y} - Y)^2]$ is also the variance of $\bar{y}$. If we cannot have $E(\bar{y}) - \bar{Y} = B = 0$ for an unbiased estimator, we want to have the bias small. For example, the ratio means $\bar{y} = y/n$ from most of the complex samples are not strictly unbiased, that is, $E(\bar{y}/n) \neq Y/N = \bar{Y}$, but the bias can be, should be, and is generally small in large samples (4.7). For most statistics from complex samples we must aim for mean square errors with small biases.

Furthermore, for most of those statistics estimates of the variances are not unbiased either: $E[\text{var}(\bar{y})] \neq \text{Var}(\bar{y}) \neq E[(y - \bar{Y})^2]$, but in large samples, well designed, we can aim at good approximations: $E[\text{var}(\bar{y})] \simeq \text{Var}(\bar{y}) \simeq E[(\bar{y} - \bar{Y})^2]$. In addition, for many statistics mathematical expressions for $\text{Var}(\bar{y})$ are unavailable and/or too complicated for practical use, but with repeated replications (BRR, jackknifing, JRR, resampling,) methods it is possible to have good and useful approximations, so that $E[\text{var}(\bar{y})] \simeq$

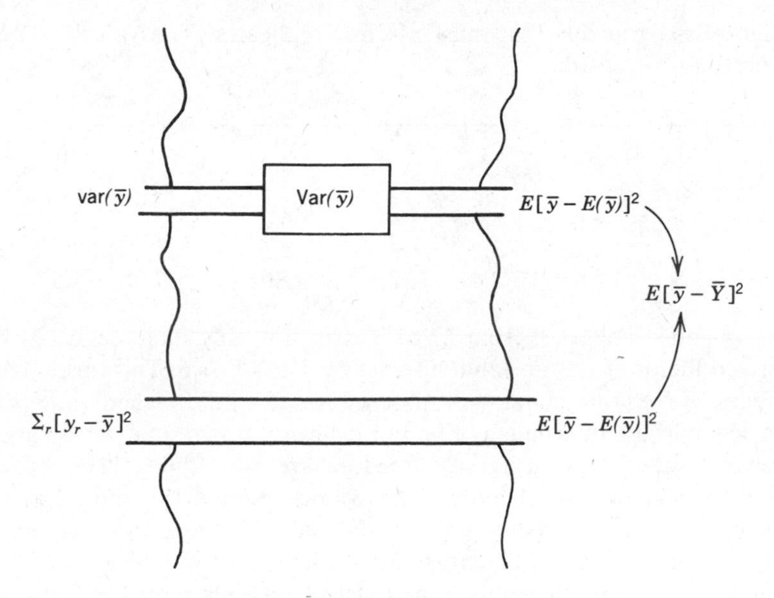

Figure 7.1.3. Repeated replication when classical methods are lacking.

Classical methods of computing variances are shown as a bridge of inference (over a river of doubts) in two sections: first the computed variance to its expectation: $E[\text{var}(\bar{y})] = \text{Var}(\bar{y})$. The second section shows that a formula for $\text{Var}(\bar{y})$ has been mathematically derived to represent $E[\bar{y} - E(\bar{y})]^2$, the variance of the statistic ($\bar{y}$) around its own expectation. There may be another short gap to the mean square error, $E(\bar{y} - \bar{Y})^2$, for biased estimates when $E(\bar{y}) - \bar{Y} = \text{bias} \neq 0$.

The lower bridge shows that resampling methods of repeated replications bridge the inference directly, without specific, mathematically derived variances. There are several methods: balanced repeated replications (BRR), jackknifes, bootstraps. They are especially needed for statistics for which mathematical derivations of variances are not available or practical. For these also there may be a gap between the expected variance and the mean square error.

$E[(\bar{y} - \bar{Y})^2]$ (7.1D) [Kish and Frankel 1974]. This point is illustrated in Figure 7.1.3.

However, more common and difficult than biases are problems of the sampling variability of the computed $\text{var}(\bar{y})$ we encounter in practice. We recall that the $E[\text{var}(\bar{y})] = \text{Var}(\bar{y})$ tends to increase with decreasing numbers of sampling units; but in addition consider that the $\text{var}(\bar{y})$ computed in any sample also varies around its expected value $E[\text{var}(\bar{y})]$. Thus in the intervals $\bar{y} \mp t_p\text{ste}(\bar{y})$, not only does $\text{ste}(\bar{y})$ increase (on the average) with decreasing size of $\sqrt{n}$, but t_p gets larger for small degrees of freedom (df).

This problem arises frequently in survey sampling, but even more in the design of experiments that must use few replicates. The values of $t_{.95}$ rise abruptly for df < 10:

df	1	2	3	4	5	6	10	30+
$t_{.99}$	63.7	9.9	5.8	4.6	4.0	3.7	3.2	2.8
$t_{.95}$	12.7	4.3	3.2	2.8	2.6	2.4	2.2	2.0
$t_{.90}$	6.3	2.9	2.4	2.1	2.0	1.9	1.8	1.7

Thus if you decrease df from 27 to 3, you not only increase ste($\bar{y}$) by an expected factor of $\sqrt{9} = 3$, but also $t_{.95}$ by $3.2/2 = 1.6$. This poses frequent conflicts for experimental and observational studies where numbers of primary units are often small (3.1). For example, if two treatments have only 3 replicates each, the degrees of freedom are only $2(3 - 1) = 4$. Theory seems to deal more with unbiased variance estimators, and neglects the problems of their instability, due to small samples. Beyond some references [Winer 1962, 5.12–5.16; Bancroft and Anderson 1974, Ch. 5; Montgomery 1984, 5.1.1] we can offer only a mere list of possible remedies for small df. Rather than defining general terms for multistage experiments, assume a comparison of two treatments given to schools × classes × students.

1. Increase the numbers of schools for both treatments.
2. Give both treatments to different classes in all schools.
3. Give both treatments to half (or part) of students in all classes.
4. Assign all students to both treatments in alternate years. "Crossover" designs.
5. "Pool" variances. This much-needed topic seems, alas, difficult to treat adequately. Pooling and averaging of variances and of design effects are practiced but seldom reported [Verma et al. 1980; Kish et al. 1976]. They are most needed for designs with internal replications restricted to only a few sites (3.1B). Searches for "similar" samples must be made. For example, repeated or periodic samples may be pooled for more degrees of freedom. Experts may help to find valid sources of data and to avoid bad mistakes.
6. Use variations within schools (either between classes or between students) for measures of error. This amounts to treating school as "fixed" effects in experimental designs (i.e., statistical inference confined to the units included in the experiment.) In sampling terms it amounts to treating a few schools as the "target" populations of case studies (3.1).

7.2 GENERALIZATIONS BEYOND THE MODULES OF 3.3

For an orderly presentation of "designs for comparisons" in Chapter 3, I developed four basic modules in 3.3. The principal features of these modules are described in 3.2A, and their limitations are admitted briefly in 3.2B. Here I intend to show that those limitations are not as narrow as they may appear on first sight. It is important to see how those modules and designs can be generalized and applied to the complex situations that researchers actually encounter and must face. For a prime example, it requires too many assumptions merely to state that comparisons are based on simple differences $(\bar{x} - \bar{y})$, with the variance simply $2\sigma^2/n$. I aim here to help the readers bridge partly the gap between those assumptions and the reality they may face.

7.2A Other Forms of Comparisons

The difference $(\bar{x} - \bar{y})$ is probably the simplest and most commonly used kind of comparison. The difference may represent two sets of individuals, two measurements on the same individuals, two subclasses of one survey sample, etc. For any difference of two random variates $\bar{x}$ and $\bar{y}$ we have:

$$\text{Var}(\bar{x} - \bar{y}) = \text{Var}(\bar{x}) + \text{Var}(\bar{y}) - 2\,\text{Cov}(\bar{x}, \bar{y}) = \sigma_{\bar{x}}^2 + \sigma_{\bar{y}}^2 - 2\sigma\bar{x}\bar{y}, \tag{7.2.1}$$

and the covariance terms always vanish whenever $\bar{x}$ and $\bar{y}$ are made independent by the design.

It seems obvious to take the next step to the difference of two differences. This may refer to the difference of two changes

$$\text{Var}[(\bar{x}_2 - \bar{x}_1) - (\bar{y}_2 - \bar{y}_1)] = \text{Var}(\bar{x}_2 - \bar{x}_1) + \text{Var}(\bar{y}_2 - \bar{y}_1) - 2\,\text{Cov}[(\bar{x}_2 - \bar{x}_1)(\bar{y}_2 - \bar{y}_1)], \tag{7.2.2}$$

or to the change in the two differences. For example, consider the male/female difference in the change of cigarette smoking between periods 1 and 2.

Often, however, the comparison may be better expressed by the ratio $(\bar{x}/\bar{y})$ than by the difference. For example, the effects of cigarette smoking are often expressed by the *ratio* of lung cancer rates of cigarette smokers over nonsmokers. The variance of the ratio can be expressed as:

$$\frac{\text{Var}(\bar{x}/\bar{y})}{(\bar{X}/\bar{Y})^2} = \frac{\text{Var}(\bar{x})}{(\bar{X})^2} + \frac{\text{Var}(\bar{y})}{(\bar{Y})^2} - \frac{2\,\text{Cov}(\bar{x}, \bar{y})}{\bar{X}\bar{Y}} =$$

$$CV^2\left(\frac{\bar{x}}{\bar{y}}\right) = C_{\bar{x}}^2 + C_{\bar{y}}^2 - 2C\bar{x}\bar{y} = \frac{\sigma_{\bar{x}}^2}{\bar{X}^2} + \frac{\sigma_{\bar{y}}^2}{\bar{Y}^2} - \frac{2\sigma\bar{x}\bar{y}}{\bar{X}\bar{Y}}. \tag{7.2.3}$$

The second line merely introduces new symbols for those on the top line, and it makes clear that the variance of a ratio is similar to the variance of a difference, except that all its terms are *relative* to the respective means of the terms. These relative variances (*relvariances*) are computed frequently and readily in sample surveys (4.7).

It is not difficult to compound the preceding two ideas in order to find expressions and variances for differences of ratios $(\bar{x}_2/\bar{x}_1 - \bar{y}_2/\bar{y}_1)$; for ratios of differences $(\bar{x}_2 - \bar{x}_1)/(\bar{y}_2 - \bar{y}_1)$; and so on. Also instead of differences we may also use ratios of ratios $(\bar{x}_2/\bar{x}_1)/\bar{y}_2/\bar{y}_1)$, also called a *double ratio*—e.g., the female/male ratio of lung cancer ratios for smokers/nonsmokers. Such ratios and linear combinations are used frequently in the construction of indexes, and their variances have been studied [Kish 1965, 12.11; Kish 1968]. We can explore their design aspects, not the analytical, computational, and mathematical issues, but it is generally feasible to approach these with modest technical resources. Conflicts in multipurpose design are likely to arise, but these may often be less severe than those due to domains (7.3).

Multivariate regression techniques may also be used for comparisons of treatments. These, however, can take so many complex forms that no brief listing would be useful here.

7.2B Assumptions for Simple Selection Designs for $(\bar{x} - \bar{y})$

Let us now unravel the assumptions that may take us from the general terms (7.2.1) for the variance of a comparison of two means to the simple $2\sigma^2/n$ for two independent means and $2(1 - R)^2/n$ for overlapping means. For the difference $d = (\bar{x} - \bar{y}) = (x/n_x) + (y/n_y)$, the general formula (7.2.1) can be written as

$$\text{Var}(d) = \text{Var}\left(\frac{x}{n_x}\right) + \text{Var}\left(\frac{y}{n_y}\right) - 2\,\text{Cov}\left(\frac{xy}{n_x n_y}\right). \tag{7.2.4}$$

If the (x/n_x) and (y/n_y) are ratio estimates, because the n_x and n_y denote variables from complex samples, then variances for differences of ratio estimates must be used (4.7.15). But when the samples n_x *and* n_y *are constants*, with no variances, then they can simply be factored out, so that

$$\text{Var}(d) = \frac{\text{Var}(x)}{n_x^2} + \frac{\text{Var}(y)}{n_y^2} - \frac{2\,\text{Cov}(x, y)}{n_x n_y}. \tag{7.2.5}$$

Now, to get to essentials, let us assume the simplest selection design for both n_x and n_y: that each was selected with simple random sampling and from large or infinite populations, so that factors $(1 - f)$ may be neglected in $\text{Var}(x) = (1 - f)n_x S_x^2$. Then $\text{Var}(x) = n_x S_x^2$ and $\text{Var}(y) = n_y S_y^2$; hence:

$$\text{Var}(d) = \frac{S_x^2}{n_x} + \frac{S_y^2}{n_y} - \frac{2n_c R_{xy} S_x S_y}{n_x n_y}. \tag{7.2.6}$$

Here R_{xy} is the correlation coefficient between the two variables, and n_c denotes the number of elements common to both of the two total samples n_x and n_y.

When the element variances are equal, $S_x^2 = S_y^2 = S^2$, then (7.2.6) becomes

$$\text{Var}(\bar{x} - \bar{y}) = S^2\left(\frac{1}{n_x} + \frac{1}{n_y} - \frac{2n_c R_{xy}}{n_x n_y}\right). \tag{7.2.7}$$

When $n_x = n_y = n$ are equal, and of these cases $n_c = P_n$ are common to both samples, then (7.2.6) becomes

$$\text{Var}(d) = \frac{1}{n}[S_x^2 + S_y^2 - 2PR_{xy}S_x S_y]. \tag{7.2.8}$$

When $S_x^2 = S_y^2 = S^2$ and $n_x = n_y = n$ are both constant, then (7.2.7) becomes the simple form we used:

$$\text{Var}(d) = \frac{2S^2}{n}[1 - PR_{xy}]. \tag{7.2.9}$$

Without an overlap $P = 0$ and $\text{Var}(d) = 2S^2/n$. With complete overlap $P = 1$ and $\text{Var}(d) = 2(1 - R_{xy})S^2/n$. Four separate cases are treated with a little more detail in [Kish 1965a, 12.4]. See also Table 6.2.3 for a comparison of four kinds of overlaps in surveys.

7.2C Unequal S_i^2, Deft_i^2, C_i, and n_i

Formulas, like $2\sigma^2/n$ for the variance of two means, often assume—implicitly or explicitly—a simple, symmetrical, uniform universe. We need simple, useful tools for dealing with the variable, unequal parameters that researchers encounter in practice. We can afford to use, as we must, rough approximations about parameters, because their errors will not affect the validity of the results, although they may reduce somewhat the efficiency of study designs. Useful designs, without great losses of efficiency, can be fashioned with rough but reasonable guesses. Errors and losses due to them will often be less than those caused by variable domains in multipurpose designs (7.3).

For the difference $(\bar{x} - \bar{y})$ of two independent samples, allocating the same size sample n to each sample produces the least variance $2S^2/n$, if (1) the element variances $S_1^2 = S_2^2 = S^2$ are equal, (2) the element costs $c_1 = c_2 = c$ are equal, and (3) the two selections are simple random (srs) though the technical, necessary conditions in (7.2.11) are looser. Otherwise, the optimal allocation of the same total size $2n$ for the two sample sizes $n_1 + n_2 = 2n$ calls for

$$n_i \propto S_i/\sqrt{c_i}, \qquad \text{that is,} \quad \frac{n_1}{n_2} = \frac{S_1/\sqrt{c_1}}{S_2/\sqrt{c_2}}. \tag{7.2.10}$$

Such "optimal allocation" produces the lowest variance for fixed total cost $\Sigma_i c_i n_i$, or the lowest cost for fixed total variance (7.3E). Differences in element variances S_i^2 and costs c_i are common, and often they can be guessed to a reasonable approximation at the time of design. But perhaps even greater differences may be found for the design effects, Deft_i^2 that have been neglected earlier. Instead of (7.2.10) we should use:

$$n_i \propto S_i\text{Deft}_i/\sqrt{c_i}, \qquad \text{that is,} \quad \frac{n_1}{n_2} = \frac{S_1\text{Deft}_1/\sqrt{c_1}}{S_2\text{Deft}_2/\sqrt{c_2}}. \tag{7.2.11}$$

The sources for differences in values of Deft_i^2 are too many and complex for even a listing here, but Sections 7.1 and 2.6 should be helpful. Using Deft^2 is merely a simplified, shortcut path into complex designs and it calls for expertise and care. Note, for example, that the Deft^2 for the difference of two means from cluster samples (2.6.3) is reduced by the covariances of the clusters; that tends to lessen the relative advantage of overlapping samples, denoted by the factor $(1 - R_{xy})$ for simple random samples.

There are several alternative ways of viewing these factors. Consider $V_i^2 = S_i^2\ \text{Deft}_i^2$ as the "effective element variance"; or consider $n_{ei} = n_i/\text{Deft}_i^2$ as the "effective element size," reduced by the design effect; and so on, to more complex factors like $Z_i = S_i^2\ \text{Deft}_i^2/\sqrt{c_i}$, or $m_i = c_i n_i/\text{Deft}_i^2$. Using these factors for comparing designs is useful even when we are vague about the parameters. Note that the sample allocation of the n_i utilizes all the parameters S_i^2, Deft_i^2, and c_i under a square root sign. For example, missing a parameter by a factor of 2 or 4 produces size misallocations of only 1.4 or 2. These in turn increase variances only by the ratios 1.04 or 1.25.

Consider, for example, a total sample size of $2n$, for a situation where $n_1 = n_2 = n$ would be the best, but $n_1 = n(1 + d)$ and $n_2 = n(1 - d)$ are used instead. Thus the variance of the comparison, instead of $1/n + 1/n = 2/n$ becomes $1/n(1 + d) + 1/n(1 - d) = 2/n(1 - d^2)$, where d is the relative reduction of one sample to compensate for the relative increase in the other. Thus a misallocation by a factor 2 would mean $(1 + d) = 4/3$ and $(1 - d) = 2/3$, and $1/(1 - d^2) = 9/8 = 1.25$. This number can be found for $K = 2$ on the top line of Table 4.5.1; see also 7.3 and [Kish 1976].

7.2D Flexibility for Sample Sizes n_i

Of the four factors involved in allocations the sample sizes n_i have the most flexibility and are most often under our control. The element variances S_i^2 are relatively inflexible and fixed, except that errors in measuring may sometimes be reduced either with better techniques or with replicated observations. The design effects Deft_i^2 may sometimes be reduced, but only with drastic changes of design, which unfortunately may increase the unit costs c_i. The c_i often turn out less easy to reduce than outsiders may naively hope.

The possibilities needed for flexibility in the numbers n_i must be pointed out, because so much in statistics and experimental design is presented in terms of fixed and symmetrical numbers n of cases. It is so much simpler to frame both theorems and illustrations that way. This limitation is also true, for example, of my Chapter 3, which I am trying to correct here. Researchers must be stimulated to overcome that rigid framework when they face the asymmetries and irregularities of the real, organic world of nature.

1. There is no need to fix exactly the sample sizes for treatments at some prespecified numbers n_i, which may be difficult in the face of missing cases, nonresponses, etc. It is even more difficult when the subjects are recruited in groups of unequal sizes, and especially when they are found in variable numbers in survey samples. Exact control under those conditions would be too expensive and potentially biasing [Kish 1977].
2. Furthermore, the numbers n_i for different treatments need not always be made the same, or even similar, even approximately. The researcher may try to approximate an optimal allocation with n_i proportional to $S_i \text{Deft}_i/\sqrt{c_i}$ (7.2C and 7.3E). Or the different n_i may be dictated by

outside constraints, especially in survey samples and in group sampling, as noted in item 1. The most obvious constraint is when only n_1 cases are available altogether for treatment 1. For example, a limited supply of a new medicine or a new technique for one treatment should not restrict the controls or the other treatments to the same numbers. The most extreme case of inequality is the One-shot Case Study [$.Ex$] in 3.4A. In this design, the control may be considered to be missing or to be the entire population.

3. On the other hand, there are also reasons for approximate equality of cases n_i between treatments. First, unequal n_i for optimal allocation of $n_i \propto S_i \, \text{Deft}_i/\sqrt{c_i}$ is only worthwhile when they differ by ratios of 2 or more, and hence the variances and costs by 4 or more (7.3.E). Second, we must remember that for a fixed variance of $\bar{x}$ in $\text{Var}(\bar{x} - \bar{y})$, increasing indefinitely the number n_y (because it is cheap) will decrease only variance of $\bar{y}$, and the other portion (perhaps half) will be unaffected. For these two very different reasons rough equality of n_i often prevails in designs.
4. Exactly fixed sample sizes n_i between treatments are also favored by two very different reasons. First, in repeated measurements on the same cases, the same numbers of elements are used on both (or all) occasions. This is the case for One-Group Pre–Post Design ($XE\tilde{x}$] in 3.4B. Second, the easy computation and mathematical elegance of fixed-size, symmetrical designs is worth some effort, when undue efforts and constraints are not against them.
5. In addition to unequal numbers of cases n_i, unequal numbers of sites (groups, units, cities) should also be considered, and they were in 3.1D.

7.2E Variation in Designs

Much of this book has been presented in terms of comparisons of two treatments: treatments versus control. Of the five basic designs in 3.4, numbers 2, 3, and 4 describe essentially two treatments, and designs 1 and 5 are close variations. However, further variations may also be desirable, and I wish merely to touch briefly on the wide range of possibilities.

First, several controls for one treatment may be used. When a new treatment is tested (a new medicine, new seed, or new method of instruction), it may compete against several current treatments. These may all be equally (un)satisfactory, with relative advantages differing over diverse situations, so that no one dominates the others over the population.

Second, several new treatments may need testing at the same time. This need can arise, for example, for a complex treatment with several separate components, when there is considerable doubt about what combination would be the most likely candidate. Then we may also consider several new treatments against several controls, with a multifactor design.

If one new treatment is tested against several controls (or vice versa), we should consider the size of the single sample (in cases and sites) against the several (k) in the control comparison. It seems reasonable to propose that the size (and expenses) for the single treatment should be greater than for each of the k treatments in the comparison, but not as great as for the k treatments combined. I assume here that the k control treatments are not simply combined into one single control (see 3.1D).

7.2F Mean Square Errors: Variances and Biases

In Chapter 3 I combined biases, variances, and costs into a single expression for each design in order to facilitate comparisons of their relative advantages. Without combining them, some designs (especially 4 and 5) would be preferred because of lower biases sometimes, whereas other designs (especially 1 and 2) would look better in others, because of lower variances (both for fixed total cost). It is common statistical practice to combine the two components of error into mean square errors = variance + bias^2, (MSE $= \sigma^2 + B^2$) and then use the root mean square error, RMSE $= \sqrt{(\text{MSE})}$, as a criterion for making inferences for probability statements, for confidence intervals, about statistical results. This concept for RMSE is basic in survey sampling, which depends heavily on inference based on the normality, through central limit theorems, of statistics based on large samples.

This rational attitude is relativistic: it considers variable errors and biases jointly as parts of the *total survey error*, which should be reduced and (it is hoped) minimized within available resources. The statistical sampling model proposes $y_{ir} = Y_i + D_{ir} = Y_i + B_r + V_{ir}$ for the rth observation on the ith element; the deviation D_{ir} of the observation y_{ir} from the true element value is $y_{ir} - Y_i = D_{ir} = B_r + V_{ir}$, separating the bias B_r from the variable error V_{ir}. "This *arbitrary* separation is the first modification toward a serviceable model, it is still too general to be an adequate frame for the concepts and measurements of empirical work." The second conceptual stage in this two-stage "dialectical" model is to recombine the separated parts into MSE $= (\Sigma_g B_g)^2 + \Sigma_v S_v^2/m_v$. "The first term is the square of the combined bias, which is the algebraic sum of all bias terms. The second term represents the sum of all variance terms, representing diverse sources, each expressed as a unit variance divided by the number m_v of these units". [Kish 1965a, 13.1–13.2]. This conceptualization should be followed by operations for measuring as many of the errors as resources allow. Those measurements should lead to procedures for reducing the errors and the MSE.

Assuming that researchers have done their best to reduce biases, the next best step is to admit their existence into their models, statistics, and inferences. "Note then that to err is human, to forgive divine—but to include errors in your design is statistical." [Kish 1978]. The next step then would be to measure biases with some precision, because this would allow researchers to adjust the results accordingly. But this can seldom be done for individuals, in the manner we can estimate our "true" heights and weights by adjusting for heels and clothes. It can be done for aggregates only now and then, and only more or less well; a few examples will help. (1) Checks against reliable outside data may be available, and these contribute to the artistic adjustments (for "turn outs" and for past biases) with which election polls are improved. (2) "Randomized response techniques" have been used to obtain aggregate answers with reduced biases [Cochran 1977, 13.17–13.18; Warner 1971]. (3) Quality checks (postenumeration surveys) have been used for censuses (seldom surveys) to obtain more accurate data. (4) More accurate measurements are sometimes obtained by taking advantages of fortunate opportunities [Ferber 1980, Kish and Lansing 1954]. However, these extra data only yield adjustments for the aggregates and means, but generally not for individual cases, hence not for relationships, regressions, etc. For noncoverage, nonresponses, especially item nonresponses, the adjustments and imputations are available only for broad classes as a rule [Kalton 1983].

We come at last to the usual and important situations where we admit to suspecting biases, but we only have vague notions about their natures and magnitudes. Theoretically the Bayesian approach would be the best for combining statistically one's guesses about biases with measures of sampling errors [Schlaifer 1983]. I fear that most researchers are seldom in a good position to apply them, even if I could teach them, but the theoretical framework is good. I propose that biases, as compared with sampling errors, are generally (or should be) smaller; are felt ("known") with less precision; and more prone to be asymmetrical. If biases were "known" with more precision, less vagueness, they could be used in adjustments. When biases are much larger than sampling errors, we should hesitate to use the sample data. Whereas sampling errors tend to be symmetrical, even approximately normal, the suspicions and fears about biases need not be so; for example, biases are not likely to be large and positive in surveys about crime, abortion, or income; each is more likely to be negative. However, usually the researcher's ideas about biases are too diffuse for the demands of Bayesian techniques.

Mean-square errors embody a fundamental idea: that we are most interested in the deviation ($\bar{y} - \bar{Y}_{\text{true}}$) of sample statistics from their "true" value. But theoretical concepts surround the subject, beginning with the squaring of the deviations before averaging:

$$\begin{aligned} \text{MSE}(\bar{y}) &= \text{Ave}[\bar{y} - \bar{Y}_{\text{true}}]^2 = \text{Ave}[\bar{y} - \text{Ave}(\bar{y})]^2 + [\text{Ave}(\bar{y}) - \bar{Y}_{\text{true}}]^2 \\ &= \sigma_{\bar{y}}^2 + \text{Bias}^2. \end{aligned} \tag{7.2.12}$$

The Bias^2 is the squared difference between two constants: the true value $\bar{Y}_{\text{true}}$ and the mean value, $\text{Ave}(\bar{y})$, of the sampling distribution of the statistic. Conceptually, the deviations must be separated into a variable sampling component σ^2 and the constant bias B^2, so that each may be treated with distinct techniques, before they are recombined in the MSE. It is a useful concept for comparing and evaluating designs, because it allows for balanced considerations of these two basic kinds of errors [Kish 1965, 13.1–13.2; Cochran 1.8–1.9]. A full theoretical justification for MSE against possible alternatives seems difficult, and some believe that the MSE is simply self-justifying with its convenience and with its squared "loss function."

For comparing designs the MSE seems to me more useful than in the analysis stage. For comparing designs we note that the error rates of $\sqrt{(\sigma^2 + B^2)}$ and of a σ' of the *same length* are rather similar for bias ratios $= B/\sigma < 1$ and even beyond. Thus using $\sqrt{(\sigma^2 + B^2)}$ is rather robust for designs. However, if in the analysis, σ is used, because the bias is not available, instead of $\sqrt{(\sigma^2 + B^2)}$, then severe understatement of errors arise for B/σ as low as 0.2 or 0.4 [Cochran 1977, Tables 1.1 and 1.2; Kish 1965a, Figs. 13.8II and III]. Of course, if B is ignored in designs, then those also suffer similar distortions.

The *relative* size of the bias, expressed by the "bias ratio" B/σ, will differ widely for the many variables and much larger number of statistics of multipurpose surveys (7.3). We may expect that for domain statistics, especially small domains, as the values of σ increase, but not the values for B, the values of B/σ will decrease as sampling errors come to dominate. Further, for differences $(\bar{y}_a - \bar{y}_b)$ of domains the biases often tend to cancel, and then sampling errors predominate as the bias ratio decreases (Figure 2.4.1) [Kish 1969].

7.3 MULTIPURPOSE DESIGNS

7.3A Purposes and Motivation

Most studies have several or many purposes present at their conception, and during the planning stages. In addition, typically many more purposes emerge later during the analysis of data and during their interpretation and utilization. "Should have many purposes" may often be more correct than "have present at their conception," because unfortunately the real multipurpose nature of many studies may lie hidden under the surface of the usual oversimplified discussions of study designs. This seems most clearly evident for sample surveys. Some surveys may even be "multisubject"; for example, data on education, economic status, and fertility may all appear jointly among the principal purposes for the same single survey. Under those principal purposes we refer only to explanatory variables and not to disturbing variables to be controlled.

While it may be true that most surveys have a single subject (e.g., fertility *or* education *or* economics as their principal purpose), several or many variables are often used to measure, describe, evaluate, and analyze that principal subject; for example, economic status may concern data on income, wealth, savings, and spending, and then each of these in turn may need measurements on several variables. Second, even a single variable (e.g., yearly income) may be described by several statistics; for example, mean income and median income, quartiles, and deciles may be presented from the same data, as well as regressions of income on other variables. Conflicts of sample design arise, because each of these statistics may benefit from different "optimal" allocation of sample sizes. Third, even more drastic conflicts arise from the diverse needs of different *domains* of analysis (2.3); such conflicts are treated in 7.3B and 7.3C.

In addition to those variables, statistics, and domains that are anticipated, planned and designed for at the outset, some other and unanticipated uses of the data are typically discovered during the analysis. "Serendipity" is common in research, and researchers should look for "side effects" of the treatments, beyond the expected responses; as in "Murphy's law," the unexpected always happens. The various side effects may be either harmful or beneficial, or both. But how can one expect and plan for the "unexpected"? Though difficult, the task is not hopeless: Answers lie in the direction of robustness and sturdiness, even at the sacrifice of fine-tuned "optimality" for only one or a few statistics (7.3C).

Multipurpose surveys are common in practice, why then are they so neglected in sampling theory? Because multipurpose theory would become too difficult and complex, and sampling theory is complex enough already. Furthermore, even the descriptions of actual sample designs tend to follow and borrow the prestige of theory, rather than to portray the many compromises of complex reality; thus those descriptions often pretend to be unipurpose. Many actual designs depart in the direction of robustness (e.g., with self-weighting, epsem samples), but explicit planning and design of multipurpose samples seems to be rare.

I discuss multipurpose design chiefly for sample surveys (which I know better), but what about other kinds of research in general? I suspect that most situations for observational studies are also multipurpose, though perhaps not as often or as strongly multipurpose as for surveys in general. As for true (ideal) experimental designs, the classical and basic theory is couched in terms of single tests of significance for rejecting a null hypothesis of zero difference between treatment means of a prespecified response variable. That situation seldom describes the real purpose and methods of actual experimental research, especially social research, I suspect.

On the other hand, it is possible that many evaluation studies (3.7) have only a single purpose, or a few related principal purposes, at least at the outset. They would often be observational studies and sometimes perhaps true experiments. But even in such studies, unexpected and side effects surface later, and then have to be considered.

Finally, the multipurpose framework has general scope; thus it can accommodate unipurpose design as merely a special case.

7.3B Areas of Conflict

Considerations for different purposes naturally lead to conflicting design specifications. First, the requirements for precision differ for various statistics. Second, sample sizes and costs may also differ, but these may be considered jointly with those precision requirements, within the same context of conflicts between purposes. Third, the parameters will also differ between statistics; for example, sampling errors (expressed probably in S^2 and Deft^2) differ between variables.

Several areas of conflict should be noted here. First, the total sample sizes n may be very different for diverse variables, and even for different statistics computed from the same variable. These differences can be much greater for domain statistics, which are often neglected but may often vie with the overall statistics for importance during analysis. For example, in a fertility survey, the rates for provinces and for age-specific rates (for single-year or five-year domains) may be as important as the overall national fertility rate, but those domain statistics would require samples ten or a hundred times larger than overall statistics, if the same relative precision were desired. Comparisons between two (or more) domain rates have even higher errors. It is often said that many samples (e.g., of size $n = 10{,}000$ or larger) are too large for the overall precision needed, especially in light of the errors of response and nonresponse, which may dominate sampling errors. However, for the domain statistics and for comparisons, the sampling errors may well dominate, because these errors become much larger, whereas nonsampling errors do not increase (Figure 2.4.1, 7.2F). Similar considerations also hold for comparisons between treatments in observational studies and experimental designs.

Second, after considering the overall sample size *n*, related questions arise about numbers of units at various stages of selection. For example, suppose about 2500 students are to be selected for each treatment. These could come from 100 classes of about 25 students each, with 4 classes from each of 25 schools (roughly); but the product of schools × classes × students has a great deal of flexibility, and considerations of costs and sampling errors vary. For sampling errors the "optimal" (preferred) considerations can vary greatly between statistics; and some estimates (or guesses) of factors and components of sampling errors and costs may be needed.

Third, researchers may want to use greater numbers of treatments for some of the explanatory variables than for others. Some treatments may be much more expensive than others, and some response variables may be especially expensive. For example, short-term responses may be obtained for all treatments, but long-term responses, requiring expensive followups, may be reserved for a subsample of them (3.1D). These conflicts may arise both in experimental designs and in observational studies.

Fourth, the allocation of sample sizes to strata (or to blocks in experiments) should depend on the characteristics of variables and of statistics. But the "optimal" (preferred) allocations may be quite different for various purposes, thus causing conflicts for the actual allocations [Kish 1965a, 3.5; Kish 1961; Kish and Anderson 1978].

7.3C Paths to Resolutions and Compromises

The topic of multipurpose design has been neglected and avoided, because it is difficult in several ways. First, the several (many, principal) purposes must be formulated explicitly in statistical terms that can also serve in the formulas for their comparisons and compromises. Obtaining such an explicit, formal, and "complete" list may be the principal obstacle. Second, estimates of the variance and cost factors are needed for each purpose. Third, values must be assigned to the required precisions for all the purposes. Fourth, from the preceding values and estimates a mathematical formulation must be created, to arrive at the solution of a single design that will be actually used. The computational tasks of such solutions have been eased by electronic computers, but the conceptual and theoretical tasks remain.

In the face of such difficulties it is no wonder that discussions of multipurpose designs are avoided in the textbooks and even in the descriptions of actual designs. Often a single statistic (e.g., the mean) of a single principal variable is presented as *the* single purpose of the study. In the framework of multipurpose design this is equivalent to assigning zero value to all other purposes. The impact of this pretense may be softened by another: that other principal purposes would result in similar allocations. But this pretense should be buttressed with calculations of the first three of the four preceding steps.

Beyond calling attention to the relevance of multipurpose design, what can I write in this nontechnical book about that technical and complex subject? Happily, we may point to two available technical approaches to the joint solution of conflicting allocations for the fourth step, after completing the first three steps.

One approach involves iterative nonlinear programming, which satisfies jointly for minimal costs the specified requirements of precisions for *all* of the stated purposes. Solutions to diverse problems have been published by several authors since about 1963 [Chatterjee 1972; Huddleston 1970; Kokan and Khan 1967]. These elegant solutions exploit the capacities of modern computers. They often come up with too high "minimal" cost, because the specified "required" precisions are often unrealistic. Then the projects "requirements" can be rescaled down to reasonably available total cost and a new solution can be recomputed. Such recomputations, however, point to the unrealistic nature of the entire procedure, which depends on precision "requirements" that usually cannot be either prespecified or fulfilled.

A very different approach calls for *averaging* between all the "optimal" (preferred) allocations for various purposes, by minimizing the combined (weighted) variance either for fixed cost or for fixed sample size. I prefer this solution, which compromises between different allocations, each of which would optimize for only one purpose. It involves assigning relative values of importance to all the listed statistics, and this may seem difficult. But the other two alternatives are more extreme, and they *should* be even more difficult: either to specify the required precisions of all statistics for the first approach, which then assigns arbitrarily equal weights of importance to all of them, or to specify one statistic for the total weight of one, and thus zero weight for all other statistics.

Furthermore, compromises are generally feasible and worthwhile, because the allocations are insensitive to moderate changes of weights (as is often true in statistics). After all, changing the relative weights (all $W_g < 1$ and $\Sigma W_g = 1$) by ratios of, for example, 2 or 5 should be less drastic than assigning the total weight 1 to one variable and 0 to all others, a process that implies infinite ratios of importance.

Three examples in 7.3D illustrate an averaging method that allows surprisingly good compromises between conflicting allocations and that is developed in detail and with references to related methods elsewhere [Kish 1976, Sections 6 and 7.6]. First, we denote with $\Sigma_i V_{gi}^2/n_i$ the variance attainable for a variate (or statistic) g with the allocation of sample sizes n_i for the ith component of variation. Then let V_g^2 (min) denote the minimal variance attainable with optimal allocation of the sample sizes, that is, with the n_i optimal for the variate g. Thus

$$1 + L_g(n_i) = (\Sigma_i V_{gi}^2/n_i)/V_g^2(\text{min}) = \Sigma_i C_{gi}^2/n_i \qquad (7.3.1)$$

may denote the ratio of increase (with the allocation n_i) in the variance of the gth variate over the minimal variance; and $L_g(n_i)$ is the relative loss over the minimal value 1. These ratios and losses will differ among various statistics g that represent the different purposes of a multipurpose survey, and those differences represent the conflicts between various allocations for any fixed total cost.

To compromise between the conflicting allocations, I propose an average loss function for any set of allocations n_i of the sample sizes, where the loss for each variate g is weighted with a factor I_g assigned for its relative importance:

$$1 + L(n_i) = \Sigma_g I_g(1 + L_g(n_i)) = \Sigma_g I_g \Sigma_i C_{gi}^2/n_i = \Sigma_i(\Sigma_g I_g C_{gi}^2)/n_i. \quad (7.3.2)$$

This may also be written as $1 + L(n_i) = \Sigma_i Z_i^2/n_i$, where $Z_i^2 = \Sigma_g I_g V_{gi}^2/V_g^2(\text{min})$, and this function may be minimized, with the sample sizes n_i optimized for the joint function, as illustrated in 7.3D.

7.3D Examples of Compromises

It should be both instructive and reassuring to see how successful some reasonable compromises can be even for the rather harsh tests posed by our examples of multipurpose designs. We may use the conflicts between designs for overall means ($\Sigma W_h\bar{y}_h$) and for separate domain means ($\bar{y}_h$) as our prime examples, because such conflicts are important and common, yet simple enough for illustration. The designs for H separate domain means $\bar{y}_h$ may be summarized by their average $\Sigma\bar{y}_h/H$. Furthermore, the designs for these averages $\Sigma\bar{y}_h/H$ may conveniently also represent designs for the comparisons $(\bar{y}_a - \bar{y}_b)$ of means, which are also important. I urge this convenient double use of $\Sigma\bar{y}_h/H$, because the variance of either the sum or the difference $(\bar{y}_a \pm \bar{y}_b)$ of two independent means is σ^2/n, if each mean is based on $2n$ independent selections (I.I.D.), and if σ^2 is the element variance for both. That variance is also σ^2/n for the average $(\bar{y}_a + \bar{y}_b)/2$ of two means, if the sample sizes are $n/2$ each, because $[\sigma^2/(n/2) + \sigma^2/(n/2)]/4 = \sigma^2/n$. Thus the averages, sums, and differences of independent sample means have the same basic forms; they differ only in the constants involving sample sizes.

To assume, for convenient simplicity, that the element variances σ^2 are similar for all means, seems disturbing. Not only may the element variances σ_h^2 differ between the separate subpopulations, but the separate sample means $\bar{y}_h$ may also be subject to different design effects D_h^2 (see 2.6), so that their variances should be expressed as $D_h^2\sigma_h^2/n_h$. Therefore we shall have to assume that σ^2 represents a proper average of the separate element variances $D_h^2\sigma_h^2$, so that we may concentrate on the effects of varying the sample sizes n_h. Because we investigate large variations in those sizes n_h, the assumption of an average value σ^2 should not be too misleading.

Another simplification is the use of sample sizes n_h to denote effort: A fixed total sample size $n = \Sigma n_h$ provides the base for comparing relative variances. However, a fixed total effort $C = \Sigma c_h n_h$ may also provide the base for comparing relative variances when element costs c_h differ between the subpopulations h; the comparisons of efficiencies for fixed C and for fixed n are similar. For example, it is possible that for large domains (large W_h in the examples below) the element costs c_h may be somewhat greater, thus "dampening" allocations proportional to the W_h (if the differences in the C_h are great enough, e.g., by factors greater than 5).

I find it more convenient and realistic to fix either the total sample size n or the total cost C and then to compare relative variances. On the other hand, alternative methods may also involve comparing relative total sample sizes or efforts C for fixed variances.

The first example concerns two subpopulations, one of which comprises the portion $W_1 = W$ of the entire population and the other of which comprises $W_2 = 1 - W$. The W can take different values, and $W = 0.5$, 0.2, 0.1, 0.01 are shown in Table 7.3.1; values of $W > 0.5$ would be redundant, because results for W and $1 - W$ are symmetrical.

The first set of results shows results for "equal allocation," that is, $n_1 = n_2$. This is optimal allocation for the average $\Sigma \bar{y}_h/H = (\bar{y}_1 + \bar{y}_2)/2$, hence also for the difference $(\bar{y}_1 - \bar{y}_2)$, as is shown by the minimal relative variance of 1 on the second line for all values of W. On the other hand, for the weighted overall average $\Sigma W_h\bar{y}_h$, the relative variances increase from 1 for $W = 0.5$ to 1.56, 1.64, 1.96 for $W = 0.2$, 0.1, 0.01. This increase has a limit of 2, as the optimal sample size $n_2 = (1 - W)n$ increases toward n instead of the allocated $n/2$.

The second set shows allocations proportional to size W, so that $n_1 = nW$ and $n_2 = n(1 - W)$. This is "optimal" for the overall mean $\Sigma W_h\bar{y}_n$ on the first line, as is shown by the minimal variance of 1 across all values of W. On the other hand, the variances for $\Sigma \bar{y}_h/H = (\bar{y}_1 + \bar{y}_2)/2$ increase from 1 for $W = 0.5$ to 1.36, 2.78, 25.25 and without limit, as $n_1 = nW$ decreases toward 0.

The relative variances for the differences $(\bar{y}_1 - \bar{y}_2)$ behave similarly to those for the sum $(\bar{y}_1 + \bar{y}_2)$ and to those for the mean $(\bar{y}_1 + \bar{y}_2)/2$ above, except for the constant $2^2 = 4$. For an extreme example, with allocation KW_h for $W = 0.01$, hence for a sample size $n_1 = W_1 n = 0.01n$, the relative variance for the mean is shown as 25.25. But the absolute variance for the difference is $4 \times 25.25\sigma^2/n = 101\sigma^2/n$, whereas the variance for the overall mean is σ^2/n. With equal allocation of $n_1 = n_2$, the relative variance for $\Sigma \bar{y}_h/H$ is shown as the minimal 1, hence the absolute variance for the difference for $(\bar{y}_1 - \bar{y}_2)$ becomes $4\sigma^2/n$, whereas the absolute variance for $\Sigma W_h \bar{y}_h$ is $1.96\ \sigma^2/n$.

The third set shows relative variances for allocations proportional to $\sqrt{W_h}$, geometric means between equal allocations and allocations proportional to W_h. These compromises are very effective: They bring large decreases in the losses for only moderate increases over the minimal values of 1.

The fourth set gives "optimal" allocations for compromising between the dual purposes of minimizing variances jointly for both $\Sigma W_h \bar{y}_h$ and $\Sigma \bar{y}_h/H$. These "optimal" allocations of n_h are proportional to $\sqrt{(W_h^2 + H^{-2})}$ and are justified elsewhere [Kish 1976, Sections 6 and 7.6]. These relative variances are generally lower than those for the other compromise with $\sqrt{W_h}$, though not always.

A better way to judge the relative efficiencies for the compromise, and indeed for all four sets of allocations, is to compare the "joint" relative variances on the third row for each of the four sets. These joint values average the two relative variances, and clearly the "optimal" compromise $\sqrt{(W_h^2 + H^{-2})}$ is better than the $\sqrt{W_h}$ compromise for each column. For the joint purposes, both compromises do considerably better than the unipurpose allocations of the first two sets. It is also clear that for very low values of W the extremely low values of $n_h = W_h n$ should be avoided.

Another interesting example concerns the populations of 133 countries, ranging in size from 0.2 to 200 millions, a range of 1000 in relative sizes. For this problem of allocation (for the World Fertility Surveys) I only omitted, for practical reasons, the four largest populations, all over 200 million, and those under 0.2 million. Their inclusion would raise the relative variance of sizes W_h from 2.5 to 12 and would make the results even more dramatic. Note that relative variances for equal sample sizes, minimal at 1 for separate country estimates, would increase to 3.34 for the global weighted average. On the other hand, proportional allocation, minimal at 1 for the global average, increases by 6.86 on the average the variances of separate country estimates. The $\sqrt{W_h}$ compromise works dramatically well for both purposes. But the statistical "optimal" allocation $\propto \sqrt{(W_h^2 + H^{-2})}$ works sensibly even better.

The "optimal" allocations have further flexibility, because they were

TABLE 7.3.1. Variance Increases (over optimal 1) for Four Allocation Methods; Two Examples

Allocations of n_h	Purpose	W: 0.5	W: 0.2	W: 0.1	W: 0.01	133 Countries
Equal	$\Sigma W_h \bar{y}_h$	1	1.56	1.64	1.96	3.34
$\propto 1/H$	$\Sigma \bar{y}_h / H$	1	1	1	1	1
	Joint	1	1.28	1.32	1.48	2.17
Proportional	$\Sigma W_h \bar{y}_h$	1	1	1	1	1
to size	$\Sigma \bar{y}_h / H$	1	1.36	2.78	25.25	6.86
$\propto W_h$	Joint	1	1.18	1.89	13.12	3.93
Square	$\Sigma W_h \bar{y}_h$	1	1.12	1.12	1.19	1.35
root	$\Sigma \bar{y}_h / H$	1	1.08	1.33	3.01	1.54
$\propto \sqrt{W_h}$	Joint	1	1.10	1.22	2.10	1.44
Optimal	$\Sigma W_h \bar{y}_h$	1	1.08	1.21	1.42	1.31
$\propto \sqrt{(W_h^2 + H^{-2})}$	$\Sigma \bar{y}_h / H$	1	1.12	1.16	1.17	1.28
	Joint	1	1.10	1.18	1.30	1.30

derived as a simple, even-handed compromise between the two specified purposes. The most obvious modification yields allocations proportional to $\sqrt{(I_c W_h^2 + I_d H^{-2})}$, where I_c and I_d are relative weights of importance ($I_c + I_d = 1$) for the combined mean and for the separate domain means. Moderate weights (I_c/I_d from 0.5 to 4) have been shown to have only slight effects [Kish 1976].

"Optimal" allocation makes interesting and good common sense on second thought, although its origin and justification are mathematical. The allocation $n_h \propto \sqrt{(W_h^2 + H^{-2})}$ has the lower limit of $1/H$ as $W_h \to 0$, and that "floor" is comforting for data based on small domains. At the other end, for large W_h it tends to be proportional to W_h and that is good for large domains. On the contrary, the geometric mean of allocation of $n_h \propto \sqrt{W_h}$ has an immediate first appeal, but it is disappointing on second thought. It goes down toward $n_h \to 0$ for small W_h, and it rises too slowly for large W_h. The advantages of "optimal" allocation show up in Table 7.3.1 in the reductions of the variance for the average domain means $\Sigma \bar{y}_h/H$, with the value 1.28 instead of 1.54 for the $\sqrt{W_h}$ allocation.

7.3E On Optimal Allocation

These brief remarks are needed because several references have been made to this topic. It is treated and indexed in all textbooks on survey sampling [e.g., Cochran 1977, 5.5, 6.14, 10.6; Kish 1965a, 8.5], but I especially recommend Kish [1976]. The treatments assume that the variance and the cost both have linear forms in several components denoted by $i(i = 1, 2, \ldots I)$, so that

$$\mathrm{Var}(\bar{y}) = \Sigma V_i^2/n_i + V_0 \quad \text{and} \quad \mathrm{Cost}(\bar{y}) = \Sigma c_i n_i + C_0. \tag{7.3.3}$$

The components may be strata or stages of sampling, and may be treatments in experimental design, etc. Note that in those linear forms the variance components are inversely proportional and the cost components directly proportional to the sample sizes n_i for the components. "Optimal allocation" refers to allocating the n_i so that either the variance or the cost is minimized with the other fixed at a required level. That is

$$VC = (\Sigma V_i^2/n_i)(\Sigma c_i n_i) \tag{7.3.4}$$

is minimized with one or the other term fixed at V_f or C_f. This results in the

$$\text{optimal } n_i \propto \sqrt{(V_i^2/c_i)} = V_i/\sqrt{c_i}, \tag{7.3.5}$$

and in either $(\Sigma V_i\sqrt{c_i})^2/C_f$ for the minimal variance or $(\Sigma V_i\sqrt{c_i})^2/V_f$ for the minimal cost. The factors of proportionality for the n_i are merely scaling factors. The simplest application is to stratified sampling where the variance terms are $\Sigma V_i^2/n_i = \Sigma W_i^2 S_i^2/n_i$, and the optimal $n_i \propto W_i S_i/\sqrt{c_i}$; i.e., the sample sizes should be proportional to the sizes and standard deviations and inversely proportional to the square root of the unit costs. Note that the square root for both variance and cost factors "dampens" the sharpness of the optimal sizes, so that moderate departures matter little.

For example, missing either cost or variance estimates by a factor of 4 misses the optimal sizes by a factor of $\sqrt{4} = 2$ only, and that increases variances only by factors between .04 to .125, as may be seen in the column for $K = 2$ in Table 4.5.1. When the c_i are relatively constant, the $n = \Sigma n_i$ is fixed or minimized and the optimal $n_i \propto V_i$.

For the difference of two means the variance is $S_a^2/n_a + S_b^2/n_b$, assuming independence for the means and for the selections; if this srs assumption does not hold, the S_i^2 may be adjusted with "design effects." Then for fixed $n = n_a + n_b$ the optimal $n_i \propto S_i$ and optimal $n_a/n_b = S_a/S_b$.

These ideas have been widely applied, amplified, and modified in the literature. We are interested in their application to a multipurpose situation, where optimal allocation is sought for a compromise variance for several (many) variates

$$\Sigma_g I_g[\Sigma_i V_{gi}^2/n_i)/V_g^2(\text{min})] = \Sigma_i[\Sigma_g I_g V_{gi}^2/V_g^2\text{min}]/n_i = \Sigma_i Z_i{}^2/n_i, \quad (7.3.2)$$

that we have seen in 7.3C. Here $Z_i{}^2 = \Sigma_g I_g V_{gi}^2/V_g^2$ (min) is the variance component to be minimized for fixed $\Sigma c_i n_i$. It is an average that uses I_g as factors of relative importance of the variates denoted by g ($\Sigma I_g = 1$). For each variate g the relative variance is given as V_{gi}^2/V_g^2 (min): the ratio of the variance with the n_i allocation to the minimal (optimal) for the same variate for the same fixed cost. Optimal allocation for the compromise leads to optimal $n_i \propto Z_i/\sqrt{c_i}$, similar to the simpler cases [Kish 1976, Section 6].

7.4 WEIGHTED MEANS: SELECTION, BIAS, VARIANCE

7.4A A Framework for Common Problems

Although weighted means are frequently needed for applications, they are neglected in modern statistical textbooks, which begin and end with identical and independent distributions (I.I.D.), where weights need not intrude; but they are treated by some older books [Yule and Kendall 1965, 14.11–14.20]. When encountering weighting problems, statisticians can often work out their own solutions from basic principles. Yet I expect that the fundamentals in 7.4C will be useful to some.

Weighting is essentially a problem in estimation, but in this book we continue to concern ourselves chiefly with its design aspects. In 7.4B we encounter three interesting problems of selection effects, each of an apparently distinct type, found isolated in the literature. But these can all be seen as having a common core in weighted means, and this is also a clue to other problems of this general kind.

Weighting is also the clue to another problem of design, treated in 7.5 as "Observational Units of Variable Sizes." Another kind of problem was dealt with in 4.5 as "Standardization: Adjustment by Weighting."

7.4B Selection Effects from Multiple Contacts, Family (Group) Members, Waiting Times

a. ***Sampling Contacts with a Facility.*** These may refer to sampling the distinct visits by persons to hospitals, libraries, theaters, stores, etc.; shares owned by shareholders; visits to doctors; etc. Consider, for example, "A Pitfall in Sampling Medical Visits" [Shepard and Neutra, 1977]:

> Samples of outpatient visits often must be used to identify users of a health facility with a given chronic condition. Such samples can lead to biases, however, because patients with more frequent visits are overrepresented. These biases can be avoided by a weighting procedure in which each sampled visit is weighted inversely to the number of clinic visits made by that patient during the sample period. This procedure proved critical in estimating the number and characteristics of hypersensitive patients seen in the medical clinic of a teaching hospital. The unweighted estimate of the number of hypertensives was 7,373 patients, more than three times the weighted estimate of 2,250. Similarly, the number of visits per year by these patients would be overestimated by almost 50 per cent without weighting. The estimated proportion of hypertensives still under treatment after 18 months was 68 per cent without weighting, compared to 51 per cent with weighting. Thus biases from failure to weight may be substantial. Analogous biases and solutions apply to other sampling problems in health services research.

They cite many other examples, and only in some of them was the selection bias corrected with weighting. They also note that, "The alternative to visit-based sampling is direct sampling of patients from a list of currently active patients at a facility."

b. ***Family Size as Selection Factor.*** This type of bias occurs often and in many forms as three examples may show:

1. "Suppose as an example that one wished to estimate the proportion of school families in a given city that have a particular characteristic—say, the proportion of school families that own their own homes. Suppose that the sample is drawn from the school records in such a way that each schoolchild has the same chance of being included.... Large families are over-represented, and smaller families under-represented.... The biases introduced by the method outlined above can be avoided or eliminated. (i) Bias can be avoided by sampling families.... (ii) Bias can be eliminated by proper weighting.... (iii) Bias can be avoided by associating family data with a unique member of the family" [Hansen, Hurwitz, Madow, I, Section 2.4B].

2. Another example from a news report from Thailand: "When the principal asked the boys and girls to tell him how many brothers and sisters they had, the children recalled the names of their brothers and sisters and counted them on their fingers. Eight or nine was an average total."

3. Preston [1976] notes that, "A widely recognized statistical fallacy is to use a value of $\bar{C}$ derived from survey responses of offspring as an estimate of $\bar{X}$, the average family size of their mothers," and that, "the mean family size of a child ($\bar{C}$) will be equal to the mean family size of women ($\bar{X}$) plus a term ($\sigma_x^2/\bar{x}$) equal to the variance of women's family sizes divided by their mean."

Preston proceeds to show sizable differences between $\bar{C}$ and $\bar{X}$ for various years in the United States.

If W_x is the proportion of women with X children, their average is simply $\bar{X} = \Sigma W_x X$. But the proportion of children who come from families of size X is $b_x = W_x X/\Sigma W_x X = W_x X/\bar{X}$. Hence the mean of these is $\bar{C} = \Sigma b_x X = \Sigma W_x X^2/\bar{X} = [\Sigma W_x(X^2 - \bar{X})^2 + \Sigma W_x \bar{X}^2]/\bar{X} = \sigma_x^2/\bar{X} + \bar{X} = \bar{X}(1 + \sigma_x^2/\bar{X}^2)$. Thus the relative increase of the biased means $\bar{C}$ over $\bar{X}$ is represented by $\sigma_x^2/\bar{X}^2 = C_x^2$, the "relvariance," or square of the coefficient of size of the families. (Or of similar groups in other selection situations.) This represents an extreme positive bias, with correlation of $+1$ between the source of the selection bias and the variable presented, both being X, the family size (number of children). This was true in examples 2 and 3; but in example 1 the variable was home ownership Y, and in this more general situation the correlation between X and Y *is* $R_{xy} < 1$, less than perfect. The biased mean is $\bar{Y}(1 + \sigma_{xy}/\bar{X}\bar{Y}) = \bar{Y}(1 + RC_x C_y)$. (7.4C)

c. ***Waiting Times.*** From a newspaper: "A committee of the American Bar Association reported in 1968 that the *average Federal prison sentence being served* in 1965 was nearly six years. . . . In contrast, *sentences* over five years are rare in most European countries." If the average length of a sentence is $\bar{X} = \Sigma W_x X$, the proportion of "sentences being served" of X years will be $b_x = W_x X/\bar{X}$ and their mean will be $\bar{C} = \bar{X} + \sigma_x^2/\bar{X}$, considerably higher than the average of sentences.

More common than leaving prison is the problem of "waiting times" for buses. Suppose you are waiting daily over the years for a London bus that starts out on schedule from the suburbs, but after delays in central London, arrives at your corner with intervals of X minutes between buses, and X may be 0, 1, 2, 3, . . ., etc. The proportion of intervals of X minutes is W_x and the mean of these is a mere $\bar{X} = \Sigma W_x X$ (similar to the mean of the bus schedule, incidentally). But the longer the interval before any one bus, the more people find that bus, so that $b_x = W_x X/\Sigma W_x X$ is the proportion of people who find the interval X. The mean of these occurrences is $\bar{C} = \Sigma b_x X = \Sigma W_x X^2/\bar{X} = \bar{X}(1 + C_x^2)$ as before, greater than $\bar{X}$ in proportion to C_x^2. The perception of the waiting customers of a much worse average than the scheduled $\bar{X}$ is correct! Incidentally, a survey at bus stops, averaging the waiting times of individual customers, would also reveal the mean $\bar{C}$ for customers, but only the mean $\bar{X}$ for buses.

At busy airports you may have noticed that many check-in counters have 0 or only 1 or 2 persons waiting, and the average length of the queues, $\bar{X}$, may be only 2 or 3. But we, the customers, usually find ourselves in the long lines, whose average length is $\bar{C} = \bar{X}(1 + C_c^2)$, which may be 10 or more.

This problem of "waiting times" or queues also receives a more mathematical treatment for continuous time and distribution theory (Poisson arrivals). The result $E(C) = E(X) + \mathrm{Var}(X)/E(X) = E(X)[1 + \sigma_x^2/E^2(X)]$ is essentially similar [Morrison 1979; Feller 1971, Eq. 4.16].

Formulation of the nature of the bias between the weighted and the unweighted mean is developed in 7.4C, point 5. A few words here may be useful for dealing with the problem. First, sometimes the problem and population may be redefined so that the weighted mean (denoted by $\bar{C}$ above and $\bar{Y}_W$ in 7.4C and 7.5) is used because it better describes the situation. We implied so much for people waiting for buses or airline counters in c, and it also may be true for some situations of family size in *b*, and some of the visits and contacts in facilities in a. In any situation it is worth raising the problem. Second, if the unweighted mean ($\bar{X}$ or $\bar{Y}_u$) is needed, it may be possible to devise means for counting the distinct persons (or other units, such as families, buses, etc.). Third, the preceding aim can sometimes be achieved with a unique counter—for example, the first visit to the hospital, or the oldest child in the family; and the others (visits, children, etc.) become blanks in the selection process. Fourth, if all persons, for example, are selected, the size of the unit must be determined for inverse weights of $1/N_\alpha$ to compensate for the selection probabilities proportional to N_α (7.5.4).

7.4C Variances and Biases in Weighted Estimates

Some fundamental formulas about weighted estimates seem desirable here, because we use them so much in this section and elsewhere. These deal with common problems in practice, which are often neglected in introductory textbooks, where all the attention is on *n* random variables, independently and identically distributed. Yet the following fundamentals can be presented simply, assuming only basic statistics from the reader [Kish 1965a, 2.8A].

1. The expectation (expected value) of a random variable (observation, event, statistic) y_j is denoted $\mathrm{Ex}(y_j)$. The expectation of y_j times a constant factor W plus a constant B (bias) is:

$$\mathrm{Exp}(Wy_j + B) = W\,\mathrm{Exp}(y_j) + B. \qquad (7.4.1)$$

The variance of $(Wy_j + B)$ is affected by the factor W but not by B:

$$\mathrm{Var}(Wy_j + B) = W^2\,\mathrm{Var}(y_j)$$
$$\text{and } \mathrm{Ste}(Wy_j + B) = |W|\mathrm{Ste}(y_j). \qquad (7.4.2)$$

The mean square error of y_j around the fixed population value Y:

$$\begin{aligned}
\text{MSE}(y_j) &= \text{Exp}(y_j - Y)^2 \\
&= \text{Exp}[y_j - \text{Exp}(y_j) + \text{Exp}(y_j) - Y]^2 \\
&= \text{Exp}[y_j - \text{Exp}(y_j)]^2 + [\text{Exp}(y_j) - Y]^2 \\
&= \text{Var}(y_j) + B^2 \qquad (7.4.3)
\end{aligned}$$

when $B = Y - \text{Exp}(y_j)$ is the bias that measures the difference between the expectation of the statistic and the population value Y; thus $B = 0$ and $\text{Exp}(y_j) = Y$ describe an "unbiased estimator" y_j of Y. The variance $\text{Var}(y_j)$ is the mean square of the deviations $\text{Exp}[y_j - \text{Exp}(y_j)]^2$ of the variable y_j around its own expectation (mean value). The capital letters for these functions denote that they represent mean values over all possible samples. The variances computed from samples should be denoted var (y_j) to show that they are variables subject to sampling errors. Then $\text{Exp}(\text{var}_j) = \text{Var}(y_j)$ denotes the situation (not always) when var_j is an unbiased estimate of the variance; but often $\text{Exp}(\text{var}_j) = \text{Var}(y_j) + \text{Bias}^2[\text{var}(y_j)]$.

2. The variance of the sum Σy_j of independent variables y_j is the sum of their variances:

$$\text{Var}(\Sigma\, y_j) = \Sigma\, \text{Var}(y_j) = \Sigma\, S_j^2. \qquad (7.4.4)$$

If the y_j are not independent we also have their covariances:

$$\text{Var}(\Sigma\, y_j) = \Sigma\, \text{Var}(y_j) + \Sigma\, \text{Cov}(y_i y_j) = \Sigma\, S_j^2 + \Sigma\, S_{ij}. \qquad (7.4.5)$$

For the sum of n variables, there are n variance terms, plus $n(n - 1)$ covariance terms that have zero expectation for independent y_j; a total of n^2 terms from the $n \times n$ matrix.

3. When the n variances are all equal $S_j^2 = S^2$ and

$$\text{Var}(\Sigma\, y_j) = nS^2 \qquad (7.4.6)$$

This is the variance of the sample sum $\Sigma\, y_j$ for an srs sample of n cases selected independently from the same identical distribution with variance S^2 for individual elements. Here we do not distinguish $\sigma^2 = S^2(N - 1)/N$, and we disregard also the factor $(1 - n/N)$ for finite population, which would give $\text{Var}\,(\Sigma\, y_j) = (1 - n/N)nS^2$ [e.g., Kish 1965a, Section 2.8].

4. Adding a constant to the variables also adds them to the expectation of their sum, so that $\text{Exp}[\Sigma\,(y_j + k_j)] = \text{Exp}(\Sigma\, y_j) + \Sigma\, k_j$; but they have no effect on the sum's variance: $\text{Var}(\Sigma\, y_j + k_j) = \Sigma\, \text{Var}(y_j)$. But multiplying by a constant factor brings the square of that factor to the variance:

$$\mathrm{Var}(\Sigma\, W_j y_j) = \Sigma\, W_j^2 \mathrm{Var}(y_j) = \Sigma\, W_j^2 S_j^2; \tag{7.4.7}$$

and if either is a constant, $W_j = W$ or $S_j = S$:

$$\mathrm{Var}(\Sigma\, W y_j) = W^2 \Sigma\, S_j^2 \quad \text{or} \quad \mathrm{Var}(\Sigma\, W_j y_j) = S^2 \Sigma\, W_j^2.$$

In particular the factor $1/n$ is used to compute the mean from the sample total: $\bar{y} = \Sigma(y_j/n) = \Sigma\, y_j/n$. Its variance is $\mathrm{Var}(\bar{y}) = \Sigma\, S_j^2/n^2$. When all $S_j^2 = S^2$ constant, as in srs, then we get the well-known

$$\mathrm{Var}(\bar{y}) = nS^2/n^2 = S^2/n. \tag{7.4.8}$$

In stratified sampling the selections are independent between strata and the weighted mean over the strata is $\Sigma\, W_h \bar{y}_h = \Sigma\, W_h y_h/n_h$, with the relative weights $\Sigma\, W_h = 1$, and often $W_h = N_h/N$, measured in population counts. Since the $W_h N_h$ and n_h (stratum sample sizes) are all fixed constants from (7.4.6) we have

$$\mathrm{Var}(\Sigma\, W_h \bar{y}_h) = \Sigma\, W_h^2 \mathrm{Var}(\bar{y}_h) = \Sigma(W_h^2/n_h^2)\,\mathrm{Var}(y_h), \tag{7.4.9}$$

and

$$= \Sigma\, W_h^2 s_h^2/n_h \qquad \text{for srs within strata.}$$

5. Differences (biases) between weighted and unweighted means $(\bar{Y}_w - \bar{Y}_u)$ pose problems in many situations (7.4B and 7.5). A simple, general expression that relates the differences to coefficients of correlation and of variation can be useful for distinguishing the likely presence of formidable from only negligible biases. Suppose that a population comprises A units and that Y_a is some value of the ath unit. From the Y_a values of all the A units one can define the usual unweighted mean

$$\bar{Y}_u = \Sigma_\alpha Y_\alpha/A. \tag{7.4.10}$$

In that unweighted mean each unit received the same constant $1/A$ as weights. On the other hand, sometimes one may want to assign variable weights and define a weighted mean

$$\bar{Y}_w = \Sigma_\alpha W_\alpha Y_\alpha = \Sigma_\alpha \frac{N_\alpha}{\Sigma N_\alpha} Y_\alpha = \frac{1}{A}\Sigma\left(\frac{N_\alpha}{\bar{N}}\right) Y_\alpha. \tag{7.4.11}$$

These three expressions illustrate the flexible nature of possible weights. The W_α are relative weights, with $\Sigma_\alpha W_\alpha = 1$. The N_α can be any arbitrary weights and we can assume that all $N_\alpha > 0$ and all $W_\alpha = N_\alpha/\Sigma N_\alpha > 0$. We also use $\bar{N} = \Sigma N_\alpha/A$, the mean weight, and to the degree that all $N_\alpha/\bar{N}$ tend to be near 1, the $N_\alpha/\Sigma N_\alpha$ tend to $1/A$, and then $\bar{Y}_w$ tends to $\bar{Y}_u$ with little difference between the two means. Variation among the N_α is a condition for a large difference (bias). The N_α can be any kind of weights for the A units, and the numbers of elements in units is one frequent use, when the units are clusters of such elements (as in 7.5). For example, the units may be cities or counties, or schools or firms, or hospitals or institutions, etc. Then our primary interest lies in values of Y_α that are mean values rather than aggregates—for example, in mean income rather than aggregate income in firms, or in rates of births and deaths per capita, rather than in total numbers of those vital events in the cities. Note that such mean unit values may be averages of element values, as well as direct measurements on the units themselves, such as the pollution or the altitude in the cities.

The difference between the two means can be viewed as

$$\bar{Y}_w - \bar{Y}_u = \frac{1}{\bar{N}}\left[\frac{1}{A}\sum_\alpha^A N_\alpha Y_\alpha - \bar{N}\bar{Y}_u\right] = \frac{1}{\bar{N}}\text{Cov}(N_\alpha, Y_\alpha). \quad (7.4.12)$$

This simply follows the definition of the covariance of the variable N_i and Y_i, similarly to definitions of variances, such as $\sigma_y^2 = (\Sigma Y_\alpha^2 - \bar{Y}_u^2)/A$ and $\sigma_n^2 = (\Sigma N_\alpha^2 - \bar{N}^2)/A$. Also, we shall find it convenient to express all of these in relative terms. The coefficients of variation are $C_y = \sigma_y/\bar{Y}_u$ and $C_n = \sigma_n/\bar{N}$. Then $\text{Cov}(N_\alpha, Y_\alpha) = R_{ny}\sigma_n\sigma_y = \bar{N}\bar{Y}_u R_{ny} C_n C_y$, from the definition of $R_{ny} = \sigma_{ny}/\sigma_n\sigma_y = \text{Cov}(N_\alpha, Y_\alpha)/\sigma_n\sigma_y$. Then the *relative difference* between weighted and unweighted means is

$$\frac{\bar{Y}_w - \bar{Y}_u}{\bar{Y}_u} = R_{ny} C_n C_y. \quad (7.4.13)$$

Thus the two means can differ emphatically when both R_{ny} and C_n are large, for any given variability C_y of the Y_α values. The weighted mean $\bar{Y}_w$ will be relatively greater (or less) than the unweighted $\bar{Y}_u$ as the weights (N_α or W_α) are positively (or negatively) correlated with the Y_α content of the A units. But a fair amount of relative variation C_n in the weights is also needed for the difference (bias) to become important. The bias can be especially large in the special cases when the $Y_\alpha = N_\alpha$ themselves, so that $R_{ny} = 1$ and the relative bias becomes $C_n C_y = C_n^2$. Of these there are examples in 7.4B and in 7.5. In 7.4B we used the symbols $\bar{X}$ for $\bar{Y}_w$ and $\bar{C}$ for $\bar{Y}_w$.

This treatment used parameters for all A units in the population, but applications to sampling of the units can be made readily, and often must be, as in 7.5. The individual values of the N_α and Y_α are assumed to be knowable constants, and it would be more complex to introduce variable measurements for them.

7.5 OBSERVATIONAL UNITS OF VARIABLE SIZES

This issue arises whenever entire groups of elements of greatly different sizes serve not only as selection units of sampling, but also as observational units. The group characteristic of each unit is observed and assigned a single score, which then can also be viewed as the value for all elements comprised within the units. First we shall note large possible differences between simple unweighted means $\bar{Y}_u$ of the units and weighted element means $\bar{Y}_w$, and hence large possible biases if the wrong mean is used. Then we shall note a method of selection (with probability proportional to size, or PPS) that eliminates that bias conveniently and efficiently. It also leads to better methods for sampling elements, as we shall see.

We emphasize the great variations of size that exist for many social groups that often are subjects of both observation and averaging. Units like cities and counties, universities and firms, hospitals and institutions can vary in size by factors of 100 or 1000 and can have coefficients or variations of size C_n of 5, 10, and more. Therefore if the correlation R_{ny} between size and the study variable is not negligible, the difference between the weighted and unweighted means can be large, because $(\bar{Y}_w - \bar{Y}_u)/\bar{Y}_u = R_{ny}C_nC_y$ represents the relative bias between them. This expression for the relative difference (bias) was developed (7.4.13) for the difference $(\bar{Y}_u - \bar{Y}_w)$ between the unweighted simple mean of units $\bar{Y}_u = \Sigma Y_\alpha/A$, and the weighted mean $\bar{Y}_w = \Sigma W_\alpha Y_\alpha = \Sigma N_\alpha Y_\alpha/\Sigma N_\alpha$. Here the weights are generally the number of elements N_α in the αth unit. The variety of examples below illustrate the pervasive nature of the phenomenon. In each of these, $\bar{Y}_u$ was first chosen automatically, but in each the mean $\bar{Y}_w$ should be considered and preferred, I propose.

1. Around 1957 (post-Sputnik) frightening statements appeared about science education in the United States: Half of the high schools offered no physics, a quarter no chemistry, and a quarter no geometry. It was later noted that, although backward schools were numerous indeed, they accounted for only 2 percent of all high school students. There were many more small schools than large schools, but the small proportion of large schools accounted for a large proportion of

students; this skewed distribution also results in a large C_n. The curricula and facilities of large and small schools differ drastically; large schools present more physics, chemistry, etc., courses than small schools; hence large R_{ny} results. Thus presenting average school conditions gives a misleading picture of conditions facing the average student. If we want the latter, the former has a large relative bias: $(\bar{Y}_w - \bar{Y}_u)/\bar{Y}_u = (0.02 - 0.50)/0.50 = -0.96$. The population elements are students, and we are interested in their opportunities. The schools serve as both sampling units and observational units of the opportunity offered by the schools to their students. (Of course the proportion of students actually taking physics, chemistry, etc., a smaller number than could pose another variable and another problem.)

2. For estimating the prevalence of swimming pools in a state's high schools, $\bar{Y}_u$ gives the proportion of schools with pools. But the weighted proportion $\bar{Y}_w$ of students enrolled in schools with pools has more meaning; and $\bar{Y}_w$ is considerably larger than $\bar{Y}_u$ because large schools have pools more often than small schools.
3. The proportion $\bar{Y}_u$ of cities with museums is not as meaningful as the proportion $\bar{Y}_w$ of people who live in (or near) cities with museums.
4. From a sample of manufacturing plants of a state, the heads were interviewed about plans to expand and to move. Instead of the simple percents $\bar{Y}_u$ of the plants, data about plants that cover $\bar{Y}_w$ percent of employees had more meaning. The two means can diverge because large and small plants can differ about their plans.
5. In some industry $\bar{Y}_u$ percent of firms operate with some defined type of organizational behavior, but a different percent $\bar{Y}_w$ of employees is subject to it. I believe $\bar{Y}_w$ is more important than $\bar{Y}_u$.
6. A national voluntary organization wants to know to what extent it is "metropolitan." Only $\bar{Y}_u$ percent of the branches are in metropolitan areas, but a much larger percent $\bar{Y}_w$ of the members belong to them, because metropolitan branches tend to be much larger.
7. In 1960, 50 million people lived in 130 U.S. cities of more than 100,000; the average size of these cities was $\bar{Y}_u = 0.39$ million, but $\bar{Y}_w = 2.0$ million was the average for inhabitants. There are many small cities, but more people live in the large cities, with the relative bias $= (2.0 - 0.39)/0.39 = 4.1$. Incidentally, using the medians (0.19 and 0.62) does not erase the difference. And of course, the contrast becomes starker if one uses a limit lower than 0.1 million.
8. The average undergraduate class size in a university was only $\bar{Y}_u = 26$, but the average student was in a class of size $\bar{Y}_w = 65$. The data of the school's statistician were grossly misleading, by the ratio $(65 - 26)/26 = 1.5$.

9. The mean number of adults per household (US, 1960) was only 2.02, but the mean number of household members was $\bar{Y}_w = 2.24$ for the average adult. This difference is small, because the dispersion (C_n) of numbers of adults is small. If we asked adults, "How many other adults live in your household?", the average would be $\bar{Y}_w - 1 = 1.24$.

The issue arises whenever the group characteristic of each unit is observed and assigned a single value, which is associated with all elements in the group. The sources of the group values may vary. The values may belong specifically to the groups, without direct source from their elements—for example, the climate, altitude, or pollution index of cities, their form of government, or the presence of museums. On the other hand, they may represent means of individual values, such as the mean income or the proportion of home owners for cities, and values arising directly from the elements, such as the population sizes and densities of the cities. From either group or individual origin, the measurements can always serve in the double capacity as group values and as values for all the elements in the groups [Kish 1965b].

We may judge that the problem exists, that it is widespread and has many varieties, and that the difference $\bar{Y}_w - \bar{Y}_u$ may be large. In many situations equal weights are assigned to the units, and values of $\bar{Y}_u$ are computed automatically and mistakenly. In my opinion, in all the preceding examples the $\bar{Y}_w$ are needed and preferable. In my experience, researchers always preferred $\bar{Y}_w$, but only *after* the difference was pointed out to them. That is the aim of this section, and to call attention to a better sample design for estimating $\bar{Y}_w$.

After choosing the proper mean come the questions about choosing preferred sample designs for estimating it. We are dealing with complete units of unequal sizes N_α, but with some special features. First, evaluation of the Y_α requires only a single observation for each unit. Second, because Y_α is the same for all N_α elements in the units, their homogeneity is extreme (rho = 1). Third, complete clusters are selected without subsampling. We now judge two alternative methods for selecting a sample of a units from a population of A units.

1. If *a units are selected with equal probability*, $f = a/A$, for all the A units, the simple mean of the sample of a values of y_α will be

$$\bar{y}_{us} = \Sigma_\alpha y_\alpha / A, \tag{7.5.1}$$

and this will be an unbiased and efficient estimate of $\bar{Y}_u$.

But to avoid a bad bias, to estimate $\bar{Y}_w$ it is necessary to weight the y_α *with* N_α:

$$\bar{y}_{ws} = \Sigma_\alpha N_\alpha y_\alpha / \Sigma_\alpha N_\alpha, \tag{7.5.2}$$

and this will be a "ratio mean" (4.7). Often the N_α can be grossly unequal, and the weights will render this estimate inefficient, with a few large units dominating the estimate and its variance (and the variance of the variance). In this situation it is more appropriate to use PPS.

2. If the *a units are selected with probabilities proportional to size* (PPS), denoted by N_α, the simple mean of the a values of y_α,

$$\bar{y}_{wp} = \Sigma_\alpha y_\alpha / a, \tag{7.5.3}$$

is (essentially) an unbiased and efficient estimate of $\bar{Y}_w$. On the other hand, a PPS sample may be used to estimate $\bar{y}_u$, with the y_α weighted with $1/N_\alpha$:

$$\bar{y}_{up} = \Sigma_\alpha (y_\alpha / N_\alpha) / \Sigma_\alpha (1/N_\alpha), \tag{7.5.4}$$

and this will also be a ratio mean, and usually not efficient.

To summarize, for these situations $\bar{Y}_w$ usually seems more meaningful, and PPS selections yield estimates that are simple and efficient. Where needed, estimates of $\bar{Y}_u$ may also be computed with weights $1/N_\alpha$. Technical details and procedures for selecting with PPS from a population of unequal-sized units may be found elsewhere [Kish 1965a, Ch. 7, Section 11.6; also 1965b]. However, we may note several points that may need attention and modification. (1) The unit variances and the unit costs may be different for large units than for small units; those differences may affect the comparisons of efficiencies for selection designs. But usually those factors will be small compared with the large effects of variation of N_α. (2) Often the exact sizes N_α are unknown and some approximate measure of size Mos_α must be used instead. Then weighting by $N_\alpha/\text{Mos}_\alpha$ may be needed for $\bar{Y}_w$, but these ratios may differ little from some constant, and the variance will be increased only slightly. (3) Stratified selection will be used often, especially for PPS selections, and that reduces the variances for both selection methods.

Interesting, special cases occur when the unit variable Y_α coincides with the unit size N_α itself. This occurs often, in preceding examples 7, 8, and 9, as answers to questions about the size of the unit to which the individual elements belong. The units are cities in example 7, university classes in 8, and households in 9. When $Y_\alpha = N_\alpha$, we have $\bar{y}_u = \bar{N} = \Sigma N_\alpha / A$ and

$$\bar{Y}_w = \frac{\Sigma N_\alpha Y_\alpha}{\Sigma N_\alpha} = \frac{\Sigma N_\alpha^2 / A}{N} = \frac{\sigma_n^2}{\bar{N}} + \bar{N},$$

and the relative difference

$$\frac{\bar{Y}_w - \bar{Y}_u}{\bar{Y}_u} = C_n^2. \tag{7.5.5}$$

This also follows as a special case of $(\bar{Y}_w - \bar{Y}_u)/\bar{Y}_u = R_{ny}C_nC_y$ (7.4.13), when $\bar{Y}_u = \bar{N}$, $R_{ny} = 1$, and $C_n = C_y$, all because $Y_\alpha = N_\alpha$. Clearly, when the relative variance C_n^2 is large, the relative difference is also great. For city size in example 7, we have $(\bar{Y}_w - \bar{Y}_u)/\bar{Y}_u = (2.0 - 0.39)/0.39 = 4.1$. However, for household size, in example 9, we have only $(2.24 - 2.02)/2.02 = 0.11$, a small relative bias. In section 7.4B, in examples c, the "waiting times" are further examples of this problem, where time spans are units of variable sizes, comprising time measures [minutes] as the elements. On the other hand, in the examples a and b of Section 7.4B, the selection method would yield $\bar{Y}_w$ automatically; hence weighting with $1/N_\alpha$ is needed to yield $\bar{Y}_u$.

Two important functions of design can be jointly fulfilled with PPS selection of groups (units) with unequal numbers N_α of individuals (elements). This occurs when, in addition to the weighted means $\bar{Y}_w$ of units, we also want a subsample of elements n_α from the sample of a units. The mean of $n = \Sigma n_\alpha$ subsampled elements $\bar{y} = \Sigma_\alpha \Sigma_\beta y_{\alpha\beta}/n$ is efficient for estimating the population mean $\bar{Y}$ of N elements (2.6). Because the group values $\bar{y}_\alpha = \Sigma_\beta y_{\alpha\beta}/n_\alpha$ are based on approximately equal numbers n_α of elements, they have approximately equal variances; and this typically enhances the statistical efficiency of the selection of unit values. These sample values $\bar{y}_\alpha$ may be related to, and analyzed together with, other unit values measured directly on the units.

Joint analyses of group variables Y_α and individual variables $X_{\alpha\beta}$ are also possible. We may regard each group value as possessing both kinds of variables, and a sample of n elements permits also their joint multivariate analysis.

7.6 ON FALSIFIABILITY IN STATISTICAL DESIGN

Statistical inference plays an important role in scientific inference. Hence statistics and statistical design cannot avoid the basic philosophical problem of empirical science: to make inferences to large populations and infinite universes and to make broad and lasting generalizations from samples of data that are limited in scope and time and that are also subject to random errors. We cannot avoid the basic philosophical questions of induction first posed clearly by David Hume in 1748.

Since then many philosophers and scientists have written on this central problem of the philosophy of science, but best and most useful (up to now) seems to be Popper's view of "falsifiability and demarcation." I have referred to "falsifiability" in several sections; a brief outline may be helpful for those not already familiar with it.

> Popper's seminal achievement has been to offer an acceptable solution to the problem of induction. In doing this he has rejected the whole orthodox view of scientific method outlined so far in this chapter and replaced it with another.
>
> Popper's solution begins by pointing to a logical asymmetry between verification and falsification. To express it in terms of the logic of statements: although no number of observation statements reporting observations of white swans allows us logically to derive the universal statement "All swans are white", one single observation statement, reporting one single observation of a black swan, allows us logically to derive the statement "Not all swans are white." In this important logical sense empirical generalizations, though not verifiable, are falsifiable. This means that scientific laws are testable in spite of being unprovable: they can be tested by systematic attempts to refute them. [Magee 1973, pp. 222–223]

Popper himself introduces his view of "Falsifiability as a Criterion of Demarcation" thus:

> The criterion of demarcation inherent in inductive logic is equivalent to the requirement that all the statements of empirical science (or all "meaningful" statements) must be capable of being finally decided, with respect to their truth *and* falsity. This means that their form must be such that *to verify them and to falsify them* must both be logically possible.
>
> Now in my view there is no such thing as induction. Thus inference to theories, from singular statements which are "verified by experience" (whatever that may mean), is logically inadmissible. Theories are, therefore, *never* empirically verifiable.
>
> But I shall certainly admit a system as empirical or scientific only if it is capable of being *tested* by experience. These considerations suggest that not the *verifiability* but the *falsifiability* of a system is to be taken as a criterion of demarcation. In other words: I shall not require of a scientific system that it shall be capable of being singled out, once and for all, in a positive sense; but I shall require that its logical form shall be such that it can be singled out, by means of empirical tests, in a negative sense: *it must be possible for an empirical scientific system to be refuted by experience.*
>
> My proposal is based upon an *asymmetry* between verifiability and falsifiability; an asymmetry which results from the logical form of universal statements. For these are never derivable from singular statements, but can be contradicted by singular statements. [Popper 1968, 6.1]

Furthermore, and most important for the advance and demarcation of good science: "What we want are statements of a high informative content, and therefore low probability, which nevertheless come close to the truth. And it is precisely such statements that scientists are interested in. The fact that they are highly falsifiable makes them also highly testable: informative content, which is in inverse proportion to probability, is in direct proportion to testability" [Magee 1973, p. 36]. Magee's small volume gives an extremely clear introduction and Salmon's book [1967] a more technical presentation of these philosophical views.

For statistical design of social research our tasks are even tougher than separating white and black swans. Our "swans" come (metaphorically) in all shades of gray from white to black and often observed through a haze. "Statistics and statisticians deal with the effects of chance events on empirical data. Because chance, randomness, and error constitute the very core of statistics, we statisticians must include chance effects in our patterns, plans, designs, and inferences" [Kish 1982]. For example, statistics must distinguish the higher risks of death from lung cancer, heart attacks, etc., for smokers, though nonsmokers are also subject to those risks and many smokers die from other causes. Furthermore, many statistical problems of research are more complex than the effects of cigarette smoking.

Many scientists and statisticians, wrestling with problems of induction, have made statements resembling parts of Popper's falsification view. Fisher's espousal of multifactor designs is an excellent example (Section 1.3); see also examples in "Strong inference" by Platt [1964]). Such parallel statements by others support Popper's view, rather than detract from his contribution. But neither in those statements nor in various philosophers' discussions could I find examples of the down-to-earth problems of statistical designs described in the following paragraphs. Yet for these problems, the principles of falsifiability present a unified view that I welcome.

A. *Internal Replication* (3.1B). If treatments and observations are introduced into several communities that are similar (adjacent) in resources, culture, organization, and administration, little additional *corroboration* is gained from the extra effort expanded on the replications. But if the replications are spread so as to increase (maximize) dissimilarities over the range of variables, over diverse conditions, over the entire signified populations of inference, then greater corroboration can be obtained from successful exposures to the severe tests of falsification. It is true that for evaluation research the range and population may have to be limited to the "dominion" of the specified program (3.7).

B. *Curves of Response Over Time* (3.6). Situations resembling A above also occur for testing the effects of treatments over time, when effects may well vary over time, but more than mere short-range effects are sought. Consider, for example, the very different effects on Romania's population growth, first one year after, then 10 years after, and 16 years after their 1966 law abolishing easy abortion.

C. *Control Strategies* (4.1). In observational studies, the difference $(\bar{y}_a - \bar{y}_b)$ between treatments A and B represents a theory about the source of that difference. But that theory must compete with others represented by uncontrolled disturbing variables. Then introducing controls over those disturbances (can the difference survive all those controls?) represents tests of falsifiability, and surviving the most severe feasible tests leads to increasing corroboration for that theory.

D. *Multifactor Experimental Designs.* R A Fisher's argument (1.3) for multifactor designs has basically much in common with those for multiple controls mentioned earlier. Yet it is even broader in that multifactor designs may begin with several factors and with interactions between them, in mutual competition for "corroboration." The basic view of severe and broad tests against falsification is similar. Note that the bold view of this scientist-statistician predated those of the philosopher, but the two converge to the same basic point, in my mind at least. Note that Fisher was also fighting against current views of induction, as Popper was. Fisher's ideas were further evolved in experimental design, especially in "response surface" analysis [Box, Hunter, and Hunter 1978, Ch.15].

Remark. The views I am about to introduce pertain as much to examples A, B, C, D as to E, F, G, H. I fear to introduce this bold modification or relation of "falsifiability," which I have not found explicitly in the philosophical presentations. However, in daily, ordinary, "normal science" and in statistical design for research, we work less often with the "falsifiability" of bold theories than with "critical estimates" (to coin a term) for important variables. For example, we estimate *how much* cigarette smoking increases death rates from lung cancer and from heart failures, by what factor and what percentage. Or we estimate the dramatic convergence toward "zero population growth" of modern industrialized nations (both Western and Eastern Europe, North America, Australia, Japan, etc.) in our generation. This concentration on estimation of important parameters is common not only in social research, but also in statistics for the biological, chemical, and physical sciences.

In statistics there is close correspondence between tests of significance and probability (confidence) intervals for estimates. Similarly, for statistical design a strong relation also exists between testing for "falsifiability" and "critical estimation" of important (crucial?) parameters. Many of the daily problems of statistical designs can be better understood in terms of estimation. Yet these can also be enriched with the falsifiability view. The problems of representation (1.1–1.4) enter here also, especially if we consider estimation not only for the entire population, but also for critical domains within it (2.3). Now we return to examples with this two-fold view in mind.

E. ***Experiments, Surveys, and Controlled Observations*** **(1.3).** This is a rich area to approach with the falsifiability view, because usually the special strength of each method—randomized treatments, representation, and realism—cannot be combined into the same research design. For example, a randomized experiment of lifetime smoking–nonsmoking habits cannot be built into a randomized selection of the U.S. population. But the effects of cigarette smoking have been shown with all three methods, including animal experiments, human surveys, and controlled groups [Cornfield 1959]. Strong corroboration results from falsifying different competing causal theories with each method, whereas the smoking–cancer link withstands the varied tests of falsification with all three methods. The Salk polio vaccine trial also involved the three-fold test [Meier 1972]. The advantages of three-fold (or two-fold) tests were already argued in Section 1.3, yet the falsifiability view seems to provide a firmer basis for the argument.

F. ***Stratification.*** Stratification is widely used in survey sampling to ensure good representation in samples of identifiable subpopulations. Reducing variances appears as the chief justification in sampling theory, and representing domains for analysis (2.3) should also be considered. But even more important in practice, I believe, is the aim of controlling for potentially disturbing variables, although this is seldom made explicit. This aim looms large, especially for large, multipurpose samples and for cases where one does not know which of many potential stratifying variables may be the most important. Such situations lead to using several, even many, stratifiers and to "multiple stratification" [Kish and Anderson 1978; Kish 1965, 12.8]. This resembles the use of controls in C above, and its basis in tests of falsification is similar, though "critical estimation" may better fit the context of survey sampling. Without stratification, the disturbing variables could affect the survey results. In experimental designs "blocking" serves similar purposes.

G. ***Randomization of Treatments*** **(1.1–1.4, 3.2).** Randomization of treatments serves, when we can use it, as the most powerful method for eliminating disturbing causal factors, which could compete with theories based on experimental treatment. Randomization provides severe tests against misleading results. After thorough investigations of social and medical innovations, Mosteller [1977a] wrote with sarcasm, "the less well controlled the study, the more enthusiasm the investigators have for the operation . . . nothing improves the performance of an innovation more than the lack of controls" [see also Gilbert, Light, and Mosteller 1975].

H. ***Representation over the Population* (1.1–1.4, 3.1).** Comparison of two methods, representative probability sampling versus internal replication (A), also illustrates the differences between tests of falsification and critical estimation. When the estimates are similar over all replications and over the entire population, both designs perform well; and internal replication may be more feasible, less costly, and more efficient. But when faced with different estimates, representative samples facilitate estimates for many kinds of domains, including those not covered by internal replications (2.3, 3.1). They also yield global averages (over all domains) that a few internal replications cannot yield, but which are often needed.

I. ***Publication of Contradictory Results.*** "On rare occasions a journal can publish two research papers back-to-back, each appearing quite sound in itself, that come to conclusions that are incomparable in whole or in part. Such a conjunction can put critical issues involving research methods and interpretations in unusually sharp focus. I recall that the *Journal* published such papers in 1978 and 1982, each with a helpful editorial. In this issue we have another such pair—and both appear to me methodologically sound. Each cites prior studies that support its conclusions and others that do not; neither is alone" [Bailar 1985].

I greet the wisdom and courage of this journal for printing both articles, but some of my friends disagree, and I invite readers to ponder and discuss. Some believe it is embarrassing for science and confusing for the public to reveal the contradiction. Should the journal wait for further evidence? But checks have already found both reports to be methodologically sound; and each is already both supported and contradicted by other studies.

Would it be better to have the two results appear in two different journals? In two different countries (such as the United States and Cuba)? On two different continents, say one in Japan, Indonesia, or Egypt? That would allow for speculations about sources of human diversity. Such differences allow conflicting results in human and social research to coexist without clear contradiction.

I believe that such contradiction is productive and at the heart of falsifiability. Out of such falsifications arise stronger theories—to be tested anew.

Problems

Professors can choose some of these problems for class assignments and examinations. There was no attempt to avoid redundancy and overlaps between questions and some are alternatives. Nor do they cover all subjects included in or related to the contents. They are mostly of the "describe" and "discuss" essay kind, mostly without unique answers. Professors may amend them with more specific targeting instructions, or may write their own versions.

Chapter 1

1a. Compare, evaluate, or rank the three major criteria of research designs. Can you suggest and justify one (or more) others? Empirical situations are welcome.

1b. Satisfying all three criteria would be best, but difficult. Describe situations (1) when that may be feasible; (2) when some specified compromise of all three seems best; and (3) when the choice of each single criterion seems preferable or necessary.

1c. Describe advantages and drawbacks for each of the major types of design in light of the three criteria (and others?), illustrated with several empirical situations.

1d. Discuss how and why representation with probability sampling may be more important and feasible either for small or for large samples.

1e. Are tests of significance applicable to survey research? Describe the controversy and add your comments.

1f. Discuss probability sampling as a sampling method and compare it with alternatives.

Chapter 2

2a. Describe and discuss the four populations of survey sampling. Also describe and discuss reasons for the three gaps between them, with one or more examples of realistic situations.

2b. Discuss effects of complex samples on descriptive statistics, on inferential statistics, and on subclasses.

2c. Discuss and compare the effects of stratification and clustering on means and other statistics.

2d. Discuss the design effects on statistics of crossclass means and their comparisons.

2e. Discuss the nature and magnitude of the effects of proportionate stratified element sampling.

2f. Discuss design effects caused by clustering on means of entire samples, on means of crossclasses, and on differences of crossclass means.

2g. Discuss real obstacles and other objections to probability sampling for analytical studies.

2h. Compare and discuss these two statements:

(a) Representing specific populations with probability sampling is important for descriptive statistics, but not for discovering fundamental relationships between variables. (Synthetic restatement of a common argument.)

(b) "... it is not true that one can uncover 'general' relationships by examining some arbitrarily selected population.... There is no such thing as a completely general relationship which is independent of population, time, and space. The extent to which a relationship is constant among different populations is an empirical question which can be resolved only by examining different populations at different times in different places" [McGinnis 1958].

Chapter 3

3a. For studies in single communities: give reasons for their frequent use, their advantages and drawbacks.

3b. Discuss reasons and designs for internal replications based on a few restricted sites.

3c. Discuss the main features and chief limitations of the four basic modules.

3d. Describe and discuss the composition of the four basic modules with their cost × variance factors and their bias type characteristics.

3e. Describe your favorite basic design among the five given or suggest a new one. You may discuss in general and/or for a defined situation.

3f. Discuss the sources and major types of bias. Suggest omissions(s) or mistake(s) in my list.

3g. Discuss the need for and problems of measuring delayed response effects.

3h. How does evaluation research (ER) differ from other research? Add your own opinions.

3i. Experts in the Treasury Department are quietly discussing the possibility of a temporary tax reduction to combat a recession, by releasing such funds into the hands of consumers. The question is how much will be spent rather than saved by consumers. It is desired to obtain some interviews on buying intentions before any plans are released to the public, and some after the tax reductions are announced and publicized widely. Discuss:

(a) Should the before and after interviews be made on the same respondents, on different respondents, or on some of each?

(b) Should the samples consist of national cross sections; be confined to a typical, homogeneous city or county; or be taken in six or eight counties, cities, or metropolitan areas, chosen to represent diverse regional, size, and economic classes?

3j. A large organization is divided into about 60 large units, separate plants and large divisions, with rather autonomous managements. It desires to conduct research on a moderate scale on the effects on production and satisfaction of a conversion from a hierarchical to a more democratic (decentralized) method of management. Discuss the advantages and drawbacks of three alternative research plans.

(a) Introduce a strong version of the new method into one entire large unit.

(b) Introduce necessarily weakened versions of the method into work sections, with matched controls, within several of the large units.

(c) In a good cross-section sample of the entire organization, conduct long and thorough interviews on the attitudes toward the possible introduction of the new methods.

Chapter 4

4a. Discuss the strategical decisions for effective control of disturbing variables.

4b. Discuss control by analysis with subclasses as to its uses, advantages, deficiencies.

4c. Describe the methods and problems of case-by-case matching.

4d. Describe alternative methods for matching subclasses.

4e. Discuss standardization as control for disturbing variables.

4f. Describe the methods and purposes of different indexes.

4g. Discuss control by covariance analysis and by residuals.

4h. Discuss the rationale and methods of adjustments with ratio estimates.

4i. Some of the adult population of 10 large cities, near which there are large atomic installations, were thought to be worried about radiation hazards. A sample of each of these cities is to be interviewed. The attitudes of the inhabitants should be contrasted with a control population, to ascertain differences in the level and prevalence of anxiety between the 10 "treatment" cities and the control population. Discuss reasons for choosing one of these alternatives as controls.

(a) A sample of the rest of the entire country.

(b) One control city, paired with each of 10 treatment cities, chosen by careful judgment as most resembling it.

(c) Three cities chosen at random from each of 10 classes of cities, each class broadly defined by size, region, etc., from the entire country and containing 1 of the 10 cities.

4j. A new vaccine against the common cold is available for a field trial of 5000 people over 10 years of age. Discuss these questions about the design:

(a) Should we use for control (1) nothing; (2) before-and-after observations on the 5000 vaccinated; (3) 5000 controls with no vaccines; (4) 5000 people with a placebo; (5) 5000 people with placebo, and before-and-after observations on all 10,000; (6) 5000 persons in a "double-blind" clinical trial with placebos; (7) any other design?

(b) Should the sample consist of (1) 5000 individually selected persons; (2) 2000 families averaging 2.5 persons each?

(c) Should the sample consist of (1) national probability sample; (2) 4 to 12 places, of contrasting characteristics on factors considered relevant by experts, such as climate, size, smoke, socioeconomic status, national origin? (3) Should these places be large cities, city tracts, suburbs, small towns, or one Army post?

(d) How many observations should be made on each individual and over how long a period?

Cost factors are (1) cost of vaccine, cost of placebo; (2) cost of reaching dwelling; (3) cost of individual observation; (4) cost of cooperation from county medical society.

Chapter 5

5a. Compare the relative advantages of samples, censuses, and registers. Give examples for defined situations. Speculate about future developments for these three methods.

5b. Discuss samples as parts of censuses for obtaining special, richer variables. Compare this method both with complete censuses and with independent samples.

5c. Describe uses of census data for sample surveys.

5d. Comment on census data as bases for separate analysis by researchers.

Chapter 6

6a. Describe your choices of reference, collection, and reporting periods for a series of objectives—e.g., births, illnesses, handicaps, income, accidents, crime.

6b. Discuss purposes in defined situations when each of the designs listed in Table 6.2.1 would serve best.

6c. Discuss a compromise design for a multipurpose situation you describe.

6d. A survey in 1955 of a large city is to be compared with another in 1988. What population changes do you expect and how do you propose to deal with them?

6e. Compare the costs, benefits, and problems of a periodic panel with distinct samples of similar sizes.

6f. Compare the costs, benefits, and problems of a periodic panel with a retrospective study over the same time interval.

6g. Discuss problems of longitudinal studies of communities, in the light of this passage from "Village in the Vaucluse" about a "typical" stable trend village of 779 people [Wylie 1964, p. 352], an anthropological and literary gem. ". . . Peyrane life persists amid all the economic, technological and architectural change: human relations in the village are about the same. . . . But of all the individuals living in Peyrane during the thirteen year period from 1946 to 1959, how many lived there for all thirteen years? Only 275! And of these, 137 were not born in Peyrane! At this point I began to wonder what I mean when I refer to 'the people of Peyrane'."

Chapter 7

7a. For each of the alternative methods for computing sampling errors describe and discuss defined situations where it seems appropriate. Describe how you would select the sample and compute sampling errors.

7b. Discuss the uses of analogies and models for imputing sampling errors from some samples to others.

7c. What statistics other than differences $(\bar{x} - \bar{y})$ can be used for comparative analysis of data?

7d. Discuss the need, possibilities, and methods for multipurpose designs.

7e. Discuss several problems of selection design that lead to biases or to weighting.

7f. Describe actual problems of organizations or groups of different sizes, problems of selection design, and solutions.

References

Numbers in **bold** type after references identify the sections where the citations occur. This serves as an index of authors and references by sections. References in books are to chapters and sections.

Agresti, A [1984] *Analyzing Ordinal Categorical Data.* New York: John Wiley & Sons. **4.6**

Althauser, R P and Wigler, M [1972] Standardization and component analysis, *Sociological Methods and Research*, **1**, 97–135. **4.5**

Alwin, D F and Sullivan, M J [1975] Issues of design and analysis in evaluation research, *Sociological Methods and Research*, **4**, 77–100. **3.7**

Anderson, D W, Kish, L, and Cornell, R G [1976] Quantifying gains from stratification for optimal strata, *JASA*, **71**, 887–892. **4.1**

Anderson, D W, Kish, L, and Cornell, R G [1980] On stratification, grouping and matching, *Scandinavian Journal of Statistics*, **7**, 61–6. **4.1**

Anderson, R L and Bancroft, T A [1952] *Statistical Theory in Research.* New York: McGraw Hill. **3.1**, **4.6**

Anderson, S, Auquier, A, Hauck, W W, Oakes, D, Vandaele, W, and Weisberg, H I [1980] *Statistical Methods for Comparative Studies.* New York: John Wiley & Sons. **4.3**

Anderson, V L and McLean, R A [1974] *Design of Experiments.* New York: Marcel Dekker. **7.2**

Andrews, F M, Morgan, J N, Sonquist, J A and Klem, L [1973] *Multiple Classification Analysis.* Ann Arbor, MI: Institute for Social Research. **4.6**

Anscombe, F J and Tukey, J W [1963] The examination and analysis of residuals, *Technometrics*, **5**, 141–60. **4.6**

Bachman, J G and Johnston, L [1978] *Monitoring the Future Project: Design and Procedures.* Ann Arbor, MI: Institute for Social Research.

Bailar, J C [1985] When research results conflict, *New England Journal of Medicine*, **313**, 1080–81. **7.6**

Berelson, B [1979] Romania's 1966 anti-abortion decree: the demographic experience of the first decade, *Population Studies*, **33**, 209–222. **3.6**

Berkson, J and Elveback, L [1960] Competing exponential risks etc., *JASA*, **55**, 415–428. **6.4**

Bickel, P J, Hammel, E A, and O'Connell, J W [1975] Sex bias in graduate admission; data from Berkeley, *Science*, **187**, 398–404. **4.2**

Bishop, Y M M, Fienberg, S E, and Holland, P M [1975] *Discrete Multivariate Analysis*. Cambridge: MIT Press. **4.6**

Blyth, C R [1972] On Simpson's paradox and the sure thing principle, *JASA*, **67**, 364–366. **4.2**

Bonham, J [1954] *The Middle Class Vote*. London: Faber. **6.6**

Boorstin, D J [1983] *The Discoverers*. New York: Random House. **3.1**

Box, G E P, Hunter, W G, and Hunter, J S [1978] *Statistics for Experimenters*. New York: John Wiley & Sons. **7.4, 7.6**

Burks, A W [1977] *Chance, Cause, Reason*. Chicago: University of Chicago Press. **1.1**

Butler, D and Stokes, D [1974] *Political Change in Britain*. London: Macmillan. **6.6**

Campbell, C and Joiner, B L [1973] How to get the answer without being sure you've asked the question, *American Statistican*, 229–237. **7.2**

Campbell, D T [1957] Factors relevant to the validity of experiments in social settings, *Psychological Bulletin*, **54**, 297–312. **3.3, 3.4, 3.5**

Campbell, D T [1969] Reforms as experiments, *American Psychologist*, **24**, 409–429. **3.6**

Campbell, D T and Stanley, J C [1963] *Experimental and Quasi-Experimental Designs*. Chicago: Rand McNally. **3.3, 3.4, 3.5, 3.6**

Cassell, C M, Sarndal, C E, and Wretman, J H [1977] *Foundations of Inference in Survey Sampling*. New York: John Wiley & Sons. **1.8**

Chatterjee, S [1972] A note on optimum stratification, *Skandinavisk Aktuaritidskrift*, **50**, 40–44. **7.3**

Cochran, W G [1957] Analysis of covariance: its nature and uses, *Biometrics*, **13**, 261–281. Entire issue devoted to covariance. **4.6**

Cochran, W G [1965] Planning of observational studies of human populations, *JRSS* (A), **128**, 234–266. **1.3, 4.1, 4.5**

Cochran, W G [1968] Errors of measurement in statistics, *Technometrics*, **10**, 637–666. **4.1**

Cochran, W G [1968] The effectiveness of adjustment by subclassification in removing bias in observational studies, *Biometrics*, **24**, 295–313. **4.6**

Cochran, W H [1969] The Use of Covariance in Observational Studies, *Applied Statistics*, **18**, 270–275. **4.6**

Cochran, W G [1977] *Sampling Techniques*. New York: John Wiley & Sons. **2.1, 2.5, 4.1, 4.7, 6.2, 7.1, 7.2, 7.3**

Cochran, W G [1983] *Planning and Analysis of Observational Studies*. New York: John Wiley & Sons. **4.1, 4.3, 4.5**

Cochran, W G and Cox, G [1957] *Experimental Designs*. New York: John Wiley & Sons. **3.1, 4.6**

Cohen, J E [1986] An uncertainty principle in demography and the unisex issues, *The American Statistician*, **40**, 32–39. **4.2**

Coleman, J S [1964] *Introduction to Mathematical Sociology*. Glencoe, IL: Free Press. **4.6**

Converse, J M [1986] *Survey Research in the US: Roots and emergence 1890–1960*. Los Angeles: University of California Press. **3.7**

Cook, D T and Campbell, D T [1976] The design and conduct of quasi-experiments and true experiments in field settings, in Dunnette, M D (ed.), *Handbook of Industrial and Organizational Research*. New York: Rand McNally. **3.5**

Cook, T D and Campbell, D T [1979] *Quasi-Experimentation: Design and Analysis Issues for Field Settings*. Chicago: Rand McNally. **3.3, 3.6**

Cornfield, J [1956] A statistical problem arising from retrospective studies, *Third Berkeley Symposium*, Vol. IV. Berkeley, University of California, 135–48. **6.4**

Cornfield, J and Haenzel, W [1960] Some aspects of retrospective studies, *Journal of Chronic Diseases*, **11**, 523–34. **6.4**

Cornfield, J, Haenszel, W, Hammond, E C, Lilienfeld, A M, Shimkin, M B, Wynder, E L et al. [1959] Smoking and lung cancer, *Journal of the National Cancer Institute*, **22**, 173–203. Also in Tufte, E R. (ed.), *The Quantitative Analysis of Social Problems*. Reading: Addison-Wesley, 1970. **4.1**

Cox, D R [1958] *Planning of Experiments*. London: John Wiley & Sons. **3.1**

Deming, W E [1943] *Statistical Adjustment of Data*. New York: John Wiley & Sons; also Dover Press. **4.4**

Dahlström, P, Jos, O, and Wahström, S [1973], The Swedish labor force surveys, *Statistical Tidskrift*, 3, 189-203. **6.2**

Deming, W E [1960] *Sample Design in Business Research*. New York: John Wiley & Sons. **7.1**

Deming, W E [1975] On probability as a basis for action, *American Statistician*, **29**, 146–152. **2.7**

Douglas, J W B and Blomfield, J M [1956] The reliability of longitudinal surveys, *Milbank Memorial Fund Quarterly*, **34**, 227–252. **4.2**

Draper, N R and Smith, H [1966] *Applied Regression Analysis*. New York: John Wiley & Sons. **4.6**

Duncan, G J [1984] *Years of Poverty, Years of Plenty*. Ann Arbor, MI: ISR Books. **6.4**

Duncan, G J and Kalton, G [1986] Issues of design and analysis of surveys across time, *International Statistical Review*, **54**. **6.1**

Duncan, G J and Mathiowetz, N A [1984] *A Validation Study of Economic Survey Data*. Ann Arbor, MI: Institute for Social Research. **7.2**

Duncan, O D and Blau, P M [1967] *The American Occupational Structure*. New York: John Wiley & Sons, 128–152. **4.6**

Edwards, W. Guttentag, M, and Snapper, K [1975] A decision-theoretic approach to evaluation research, in Struening, E L and Guttentag, M, *Handbook of Evaluation Research*. Beverly Hills, CA: Sage Publications. **3.7**

Efron, B [1982] *The Jackknife, the Bootstrap and Other Resampling Plans*. Philadelphia: SIAM. **7.1**

Ericson, W A [1969] Subjective Bayesian models in sampling finite populations, *JRSS*(B), **31**, 195–233. **1.8**

Feller, W [1971] *An Introduction to Probability Theory and Its Applications*, Vol. II. New York: John Wiley & Sons. **7.4**

Ferber, R [1980] *What is a Survey*? Washington: American Statistical Association. **1.1**

Fienberg, S E [1977] *The Analysis of Cross-Classified Categorical Data*. Cambridge: MIT Press. **4.2**

Fienberg, S F and Mason, W M [1979] Identification and estimation of age-period-cohort models, in Schuessler, K F (ed.), *Sociological Methodology 1979*. San Francisco: Jossey-Bass, 1–67. **6.3**

Fienberg, S E, Singer, B, and Tanur, S M [1985] Large-scale social experimentation in the United States, Ch 12 in Atkinson, A C and Fienberg, S E, *A Celebration of Statistics*, New York: Spoinger Verlag. **3.7**

Fisher, R A [1935] *The Design of Experiments*. 1st ed. London: Oliver and Boyd. **1.1**

Fleiss, J L [1973] *Statistical Methods for Rates and Proportions*. New York: John Wiley & Sons. **4.3**

Forthofer, R N and Lehnen, R G [1981] *Public Program Analysis: A New Categorical Data Approach*. Belmont, CA: Lifetime Learning Publications. **4.6**

Freedman, D S, Thornton, A. Camburn, D [1980] Maintaining response rates in longitudinal studies, *Sociological Methods and Research*, **9**, 87–98. **4.2, 6.4**

Freedman, R and Hawley, A H [1949] Unemployment and migration in the depression (1930–1935), *JASA*, **44**, 260–272. **4.3**

Galtung, J [1975] Diachronic analysis, structure and change, in Blalock H M et al. (eds.), *Quantitative Sociology*. New York: Academic Press. **3.6**

Gilbert, J P, Light, R J, and Mosteller, F [1975] Assessing social innovations, in Bennett, C A and Lumsdane, A A (eds.), *Evaluation and Experiment*. New York: Academic Press. **3.5, 6.6, 7.6**

Glass, G V [1976] Primary, secondary and meta-analysis of research, *Educational Researcher*, **5**, 3–8. **6.6**

Glass, G V, McGaw, B, and Smith, M L [1981] *Meta-Analysis in Social Research*, Beverly Hills, CA: Sage Publications. **6.6**

Goldstein, H [1979] *The Design and Analysis of Longitudinal Studies*. London: Academic Press. **3.6, 6.1**

Gonzalez, M, Ogus, J L, Shapiro, G, and Tepping, B J [1975] Standards for discussion and presentation of errors in survey and census data, *JASA*, **70** (351), part II, **7.1**

Gray, P G [1955] The memory factor in social surveys, *JASA*, **50**, 344–363. **6.1**

Greenberg, B J [1969] Problems of statistical inference in health, *JASA*, **64**, 739–358. **6.4**

Groves, R G [1987] *Survey Errors and Costs*, forthcoming. **7.2**

Groves, R M and Kahn, R L [1979] *Surveys by Telephone*, New York: Academic Press, Secs. 6.3–6.5. **7.1**

Haas, D H [1982] Survey sampling and the logic of inference in sociology, *The American Sociologist*, **17**, 103–111. **1.8**

Hansen, M H, Hurwitz, W N, and Bershad, M A [1961] Measurement errors in censuses and surveys, *Bulletin of the International Statistical Instute*, **38**, 359–374. **5.2**

Hansen, M H, Hurwitz, W N, and Madow, W G [1953] *Sample Survey Methods and Theory*, Vol I. New York: John Wiley & Sons. **2.1, 2.6, 4.7, 7.1, 7.4**

Hansen, M H, Madow, W S, and Tepping, B J [1983] An evaluation of model dependent and probability sampling inferences in sample surveys, *JASA*, **78**, 776–807. **1.8**

Harris, C W (ed.) [1963] *Problems in Measuring Change*. Madison: University of Wisconsin Press. **6.1**

Harrop, M, England, J, and Husbands, C T [1980] The bases of national front support, *Political Studies*, **28**, 271–283. **6.6**

Hedges, L V and Olkin, I [1985] *Statistical Methods for Meta-Analysis*. Orlando, FL: Academic Press. **6.6**

Herringa, S G [1981] *Small Area Estimation: A Review*. Ann Arbor, MI: Institute for Social Research. **5.3**

Herriott, R A and Kasprzyk, D [1984] The survey of income and program participation (SIPP), *Proceedings of the Section on Social Statistics, American Statistical Association*. **6.2, 6.3**

Hess, I [1985] *Sampling for Social Research Surveys, 1974–80*. Ann Arbor, MI: Institute for Social Research, Ch. 3. **6.2**

Hess, I, Riedel, D C, Fitzpatrick, T B [1975] *Probability Sampling of Hospitals and Patients.* Ann Arbor, MI: Health Administration Press, chap. VII. **7.2**

Hill, B A [1961] *Principles of Medical Statistics.* London: The Lancet. **4.5**

Hoaglin, D C, Light, R J, McPeek, B, Mosteller, F, and Stoto, M A [1982] *Data for Decisions.* Cambridge, MA: Abt Books. **3.7**

Holt, D, Scott, A J, and Ewings, P D [1980] Chi-squared tests with survey data, *JRSS* (A), **143**, 303–20. **7.1**

Holt, D, Smith, T M F, and Winter, D [1982] Regression analysis of data from complex surveys, *JRSS* (A), **143**, 474–487. **7.1**

Horvitz, D G and Thompson, D J [1952] A generalization of sampling without replacement from a finite universe, *JASA*, **47**, 663–85. **7.1**

Huddleston, H F, Claypool, P L, and Hocking, R R [1970] Optimal sample allocation to strata using convex programming, *Applied Statistics*, **19**, 273–278. **7.3**

Hyman, H H [1955] *Survey Design and Analysis.* Glencoe, IL: The Free Press. **4.2**

Janson, C G [1981] Some problems of longitudinal research in the social sciences, in Schulsinger et al. **6.1, 6.4**

Janson, C G [1984] *Project Metropolitan 21*, University of Stockholm, Sociology Dept. Research Report 21. **4.2, 6.1**

Jessen, R J [1978] *Statistical Survey Techniques.* New York: Wiley-Interscience. **1.4**

Juster, F T [1985] The validity and quality of time-use estimates obtained from recalls, *JASA*, **80**. **3.5, 6.4**

Kahn, R L [1975] In search of the Hawthorne effect, in Cass, E L and Zimmer, F G (eds.), *Man and Work in Society*, New York: Van Nostrand Reinhold. **1.3, 3.5, 3.6**

Kalton, G [1968] Standardization: a technique to control for extraneous variables, *Applied Statistics*, **17**, 118–136. **4.5**

Kalton, G [1979] Ultimate cluster sampling, *JRSS* (A), **142**, 210–22. **2.6**

Kalton, G [1983] *Compensating for Missing Data.* Ann Arbor, MI: Institute for Social Research. **4.3, 7.2**

Kalton, G [1983] Models in the practice of survey sampling, *International Statistical Review*, **51**, 175–188. **1.8**

Kalton, G and Kish, L [1981] Two efficient random imputation procedures, *Proceedings of Section on Survey Research Methods, American Statistical Association.* **4.3**

Kannisto, V [1983] *Collection of Migration Data Through a Follow-up Survey.* University of Louvain: Demography Department. **6.4**

Kendall, M G and Buckland, W R [1971] *A Dictionary of Statistical Terms*, 3rd ed., London: Longman's. **7.1**

Kendall, M G and Stuart, A [1966] *The Advanced Theory of Statistics.* London: Griffin and Co. **7.1**

Kessler, R C and Greenberg, D F [1981] *Linear Panel Analysis.* New York: Academic Press. **6.2**

Keyfitz, N [1953] A factorial arrangement of comparisons of family size, *American Journal of Sociology*, **53**, 470. **4.5**

Kirk, R E [1968] *Experimental Design: Procedures for the Behavioral Sciences.* Belmont, CA: Wadsworth. **4.6**

Kish, L [1957] Confidence intervals, for complex samples, *American Sociological Review*, **22**, 165–165. **7.2**

Kish, L [1959] Some statistical problems in research design, *American Sociological Review*, **24**, 328–338. **1.3, 1.5**

Kish, L [1965a] *Survey Sampling*. New York: John Wiley & Sons. **1.4, 1.7, 2.1, 2.2, 2.3, 2.5, 2.6, 3.1, 3.4, 4.1, 4.3, 4.5, 6.1, 6.2, 6.3, 6.6, 7.1, 7.2, 7.3, 7.4, 7.5, 7.6**

Kish, L [1965b] Sampling organizations and groups of unequal sizes, *American Sociological Review*, **20**, 564–572. **7.5**

Kish, L [1968] Standard errors for indexes from complex samples, *JASA*, **63**, 512–529. **7.2**

Kish, L [1969] Design and estimation for subclasses, comparisons and analytical statistics, in Johnson, N L and Smith, H (eds.) *New Developments in Survey Sampling*. New York: John Wiley & Sons. **2.5, 7.2**

Kish, L [1975] Representation, randomization and control, in Blalock, H M, Aganbegian, Borodkin, Bodin, and Capecchi (eds.), *Quantitative Sociology*. New York: Academic Press. **1.3**

Kish, L [1976] Optima and proxima in linear sample designs, *JRSS* (A), **139**, 80–95. **7.6, 4.5**

Kish, L [1977] Robustness in survey sampling, *Bulletin of the International Statistical Institute*, **47** (3), 515—28. **7.1**

Kish, L [1979] Samples and censuses, *International Statistical Review*, **47**, 99–109. **5.1, 5.2, 5.3**

Kish, L [1980] Design and estimation for domains, *The Statistician* (London Institute of Statisticians), **29**, 209-222. **2.3, 2.5**

Kish, L [1981] *Using Cumulated Rolling Samples*. Washington: Library of Congress, US Printing Office No. 80-528-0. **6.6**

Kish, L [1984] On analytical statistics from complex samples. *Survey Methodology* (Statistics Canada, Ottawa), **10**, 1, 1–7. **7.1**

Kish, L [1985] Sample surveys versus experiments, controlled observations, censuses, registers and local studies, *Australian Journal of Statistics*, **27**, 111–122. **1.3**

Kish, L [1986], Timing of surveys for public policy, *Austrian Journal of Statistics*, 28, 1-12. **6.2**

Kish, L and Anderson, D W [1978] Multivariate and multipurpose stratification, *JASA*, **63**, 1298–1309. **3.1, 4.1, 7.3, 7.6**

Kish, L and Frankel, M R [1974] Inference from complex samples, *JRSS* (B), **36**, 1–37. **2.2, 2.6, 7.1**

Kish, L, Groves, R M, and Krotki, K P [1976] *Sampling Errors for Fertility Surveys*. The Hague: International Statistical Institute. **2.6, 7.1**

Kish, L and Lansing, J B [1954] Response errors in estimating the values of homes, *JASA*, **49**, 520–538. **7.2**

Kish, L and Verma, V [1983] Censuses and samples: combined uses and designs, Bulletin of the International Statistical Institute, **44**, Book 1, 66–82, **5.1, 5.3**

Kish, L [1987] Multipurpose sample design, *International Statistical Review*, (submitted). **7.3**

Kish, L, Lovejoy, W and Rackow, P [1961] A multi-stage probability sample for continuous traffic surveys, *Proceedings of the Social Statistics Section, American Statistical Association*, 66–70. **6.2**

Kitagawa, E M [1955] Components of a difference between two rates, *JASA*, **50**, 1168–1194. **4.5**

Kitagawa, E M [1964] Standardized comparisons in population research, *Demography*, 296–315. **4.5**

Kleinbaum, D G, Kupper, L L and Morgenstern, H [1982] *Epidemiological Research*. Belmont, CA: Lifetime Learning Publications. **2.7, 4.3**

Kokan, A R and Khan, S [1967] Optimum allocation in multivariate surveys; an analytical solution, *JRSS* (B), **29**, 115–125. **7.3**

Kotz, S and Johnson, N L [1982] *Encyclopedia of Statistical Sciences*. New York: Wiley-Interscience.

Kruskal, W and Mosteller, F [1979–80] Representative sampling, I, II, III, and IV, *International Statistical Review*, **47**, **48**. **1.7**

Kruskal, W H and Tanur, J M [1978] *International Encyclopedia of Statistics*. New York: The Free Press. **7.2**

Landis, J R, Heyman, E R, and Koch, G G [1978] Average partial association in three-way contingency tables: a review and discussion of alternative tests, *International Statistical Review*, **46**, 237–254. **7.1**, **4.6**

Landis, J R, Lepkowski, J M, Eklund, S A, and Stehouwer, S H [1982] *A Statistical Methodology for Analyzing Data from a Complex Survey*. Washington: National Center for Health Statistics, Series 2, No. 92. **7.1**

Lepkowski, J M and Landis, J R [1982] Design effects for linear contrasts of subclass proportions, *International Statistical Review*. **7.1**

Light, R J (ed.) [1983] *Evaluation Studies Review Annual*, Vol. 8, Beverly Hills, CA: Sage Publications. **6.6**

Madow, W G, Olkin, I and Rubin, D B [1983] *Incomplete Data in Sample Surveys*, 3 vols. New York: Academic Press. **2.1**

Magee, B [1973] *Popper*. London: Fontana Paperbacks. **3.1**, **7.6**

Mantel, N and Haenzel, W [1959] Statistical aspects of the analysis of data from retrospective studies of disease, *Journal of the National Cancer Institute*, **22**, 719–748. **6.4**

Marks, E S, Seltzer, W, and Krotki, K K [1974] *Population Growth Estimation*. New York: The Population Council. **5.3**, **6.1**

Marshall, W A and Swan, A [1971] Seasonal variation in growth rates of normal and blind children, *Human Biology*, **43**, 502–516. **3.6**

McGinnis, R [1958] Randomization inference in sociological research, *American Sociological Review*, **23**, 408–414. **1.5**, **2.7**

McKinley, S [1975] The design and analysis of the observational study–a review, *JASA*, **70**, 503–523. **1.3**, **4.1**

McMillen, D B and Herriott, R A [1984] Toward a longitudinal definition of households, *Proceedings of the Section on Social Statistics, American Statistical Association*. **6.1**

Mednick, S A, Baert, A E, Bachman, B P (eds.) [1981] *Prospective Longitudinal Research* (Psychosocial Research for WHO). London: Oxford University Press. **6.4**

Meier, P [1972] The biggest public health experiment ever: the 1954 field trial of the Salk polio vaccine, in Tanur, J M et al. (eds.), *Statistics: Guide to the Unknown*. San Francisco: Holden-Day. **1.4**, **3.5**

Miller, W L [1983] *The Survey Method in the Social and Political Sciences*. New York: St. Martin's Press. **6.6**

Montgomery, D C [1984] *Design and Analysis of Experiments*. New York: John Wiley & Sons. **7.2**

Morgan, J N, David, M H, Cohen, W J and Brazer, H E [1962] *Income and Welfare in the United States*. New York: McGraw-Hill. **4.6**

Morrison, D G [1973] Some results for waiting times with an application to survey data, *American Statistician*, **27**, 226–227. **7.4**

Moser, C A and Kalton, G [1971] *Survey Methods in Social Investigations*. London: Heinemann. **4.2**

Moss, L and Goldstein, H (eds.) [1979] *The Recall Method in Social Surveys*. London: University of London, Institute of Education. **6.4**

Mosteller, F [1967] Nonsampling errors, in the *International Encyclopedia of the Social Sciences*. New York: Macmillan. **3.6**, **7.2**

Mosteller, F [1977] Experimentation and innovations, *Bulletin of the International Statistical Institute, 41st Session*. **4.1**, **7.6**

Mueller, J H, Schuessler, K F, and Costner, H L [1970] *Statistical Reasoning in Sociology*. Boston: Houghton Mifflin. **4.5**

Namboodiri, N K [1970] A Statistical Exposition of the "before-after" and "after-only" designs, *American Journal of Sociology*, **76**, 83–102. **3.3**, **3.4**, **3.5**

National Academy of Sciences [1981] *Collecting Data for the Estimation of Fertility and Mortality*. Washington: National Academy Press. **6.1**

National Center for Health Statistics [1958] *Statistical Designs of the Health Household—Interview Survey*, Public Health Series 584-A2, 15–18. **6.4**, **6.6**

National Center for Health Statistics [1980] *The Person-Number System etc*. Washington: NCHS Series 2, No. 84. **5.3**

Neter, J and Wasserman, W [1974] *Applied Linear Statistical Models*. Homewood IL: Richard D Irwin. **4.6**

Neyman, J [1934] On the two different aspects of the representative method, *JRSS*, **97**, 558–625. **1.1**, **1.4**

Neyman, J [1952] *Lectures and Conferences on Mathematical Statistics and Probability*. Washington: Graduate School of USDA. **1.1**

O'Muircheartaigh, C and Payne, C [1977] *The Analysis of Survey Data*. 2 vols. New York: John Wiley & Sons. **4.2**

O'Muircheartaigh, C and Wong, S T [1981] The impact of sampling theory on survey sampling practice: a review, *Bulletin International Statistical Institute*, 43rd Session I, 465–493. **3.1**

Platt, J [1964] Strong inference, *Science*, **146**, 347–353. **7.6**

Plewis, I [1985] *Analyzing Change: Measurement and Explanation Using Longitudinal Data*. New York: John Wiley & Sons. **6.1**

Popper, K R [1959] *The Logic of Scientific Discovery*. London: Hutchinson (esp. Chs. 1, 10). **1.1**, **3.1**, **7.6**

Preston, S [1976] Family sizes of children and family sizes of women, *Demography*, **13**, 105. **7.4**

Purcell, N R and Kish, L [1979] Estimation for small domains, *Biometrics*, **35**, 365–384. **5.3**

Purcell, N J and Kish, L [1980] Postcensal estimates for local areas (small domains), *International Statistical Review*, **48**, 3–18. **2.3**, **5.3**

Radner, D B [1979] The development of statistical matching in economics, *Proceedings of the Social Statistics Section, American Statistical Association*, 503–508. **4.3**

Radner, D B [1980] *Exact and Statistical Matching Techniques*. Washington: US Dept of Commerce, Statistical Working Paper No. 5. **4.3**, **4.4**

Rao, C R [1985] Weighted distributions arising out of methods of ascertainment, Ch. 24 in Atkinson, A C and Fienberg, S E, *A Celebration of Statistics*. New York: Springer Verlag. **7.4**

Rao, J N K and Wu, C F J [1984] Bootstrap inference for sample surveys, *Proceedings of Section on Survey Research Methods, American Statistical Association*, 106–12. **7.1**

Rao, J N K, Platek, R, Sarndal, C E, and Singh, M P (eds.) [1986], *Small Area Statistics: An International Symposium*. New York: Wiley-Interscience. **5.3**

Rodgers, W L [1982] Estimable functions of age, period and cohort effects, *American Sociological Review*, **47**, 774–796. **4.3, 4.4**

Rosenbaum, P R [1984] The consequences of adjustment for a concomitant variable that has been affected by the treatment, *JRSS* (A), **147**, 656–666. **4.1**

Rosenbaum, P R and Rubin, D B [1985] Constructing a control group etc, *The American Statistician*, **39**, 33–38. **4.1**

Rosenberg, M [1968] *The Logic of Survey Analysis*. New York: Basic Books. **4.2**

Ross, J and Smith, P [1968] Orthodox experimental designs, Ch. 9 in Blalock, H M and A B (eds.), *Methodology in Social Research*. New York: McGraw-Hill. **3.3, 3.4, 3.5**

Rossi, P H and Freeman, H E [1985] *Evaluation: A Systematic Approach*, 3rd ed. Beverly Hills, CA: Sage Publications. **3.7**

Rubin, D B [1978] Multiple imputations in sample surveys, *Proceedings of Section on Survey Research Methods, American Statistical Association*. **4.3**

Rust, K F [1984] *Techniques for Estimating Variances for Sampling Surveys*, Ph.D. dissertation, Ann Arbor: University of Michigan. **7.1**

Rust, K F [1985] Variance estimation for complex estimators in sample surveys, *Journal of Official Statistics*, **1**, 381–398. **7.1**

Sacks, O [1973] *Awakenings*. New York: Dutton. **3.6**

Salmon, W C [1967] *The Foundation of Scientific Inference*. Pittsburgh: University of Pittsburgh Press. (esp. Secs. 1, 7). **1.1, 3.1, 7.6**

Schlaifer, R [1983] *Analysis of Decisions under Uncertainty*. Malabar, FL: Krieger. **7.3**

Schlesselmann, J J and Stolley, P D [1982] *Case-Control Studies: Design, Conduct, Analysis*. New York: Oxford University Press. **4.3**

Schulsinger, F, Mednick, S A, and Knop, J (eds.) [1981] *Longitudinal Research: Methods and Uses in Behavioral Science*. Boston: Martinus Nijhoff. **6.1**

Schuman, H and Kalton, G [1985] Survey methods, Ch. 12 in Lindzey, S and Aronson E, (eds.) *The Handbook of Social Psychology*. 3rd ed. New York: Random House. **4.2**

Selvin, H C [1957] A critique of tests of significance in survey research, *American Sociological Review*, **22**, 527. **1.5**

Shepard, D S and Neutra, R [1972] A pitfall in sampling medical visits, *Am. J. of Public Health*, **67**, 743–750. **7.4**

Shryock, H S and Siegel, J S [1973] *The Methods and Materials of Demography*. Washington: US Bureau of the Census, esp. 418–423, 481–486. **4.5**

Sikkel, D [1985] Models for memory effects, *JASA*, **80**, 835–844. **6.1**

Simpson, E H [1951] The interpretation of interaction in contingency tables, *JRSS* (B) **13**, 238–41. **7.4**

Snedecor, G W and Cochran, W G [1967] *Statistical Methods*. Ames: Iowa State University Press. **3.1, 4.6**

Speed, F M, Hocking, R R and Hackney, O P [1978] Methods of analysis of linear models with unbalanced data, *JASA*, **73**, 105–112. **7.4**

Statistics Canada [1984] Selected bibliography of data analysis for complex samples, *Survey Methodology* (Ottawa), **10**, 1, 119–125. **7.1**

Struening, E L and Guttentag, M [1965] *Handbook of Evaluation Research*. Beverly Hills, CA: Sage Publications. **3.7**

Suchman, E A [1962] *Evaluation Research*. New York: Russell Sage. **3.7**

Sudman, S and Bradburn, N M [1973] Effects of time and memory factors on response in surveys, *JASA*, **68**, 805–815. **6.1**

Sudman, S and Bradburn, N [1984] *Response Effects in Surveys*. Chicago: Aldine. **7.2**

Suits, D E [1957], Dummy variables in regression equations, *JASA*, 66, 548-551. **46**

Sukhatme, P V [1947] The problems of plot size in large-scale surveys, *JASA*, **42**, 297–310. **1.4**

Sukhatme, P V [1954] *Sampling Theory of Surveys with Applications*. Ames: Iowa State College Press. **1.4**

Susser, M [1972] *Causal Thinking in the Health Sciences*. New York: Oxford University Press. **1.4**

Susser, M [1975] Epidemiological Models, Ch. 14 in Struening, E L and Guttentag, M (eds.), *Handbook of Evaluation Research*. Beverly Hills, CA: Sage Publications. **3.7**

Swicegood, C G, Morgan, S P, and Rindfuss, R R [1984] Measurement and replication: evaluating the consistency of eight US fertility surveys, *Demography*, **21**, 19–33. **6.6**

Tam, S M [1984] On covariances from overlapping samples, *American Statistician*, **38**, 288–289.

Teitelbaum, M S [1972] Fertility effects of the abolition of legal abortion in Romania, *Population Studies*, **26**, 405–18. **3.6**

Thornton, A, Freedman, D S, and Camburn, D [1982] Obtaining respondent cooperation in family panel studies, *Sociological Methods and Research*, **11**, 33–51. **6.4**

Tufte, E R (ed.) [1970] *The Quantitative Analysis of Social Problems*. Reading, MA: Addison-Wesley. **1.1**, **3.6**

Turner, C F and Martin, E [1985] *Survey Subjective Phenomena*, Vols. 1 and 2. New York: Russell Sage Foundation. **7.2**

United Nations [1962] *Methodology and Evaluation of Continuous Population Registers*. New York: United Nations. **5.3**

United Nations [1977] Study of special techniques for enumerating nomads in African Censuses and Surveys, ECAFE:E/CN.14/CAS 10/16. **6.3**

United Nations [1984] *The Follow-up Approach: A Method for Measuring Natality, Mortality and Migration*. UN Statistical Office. **6.4**

United Nations Statistical Office [1980] Principles and recommendations for population and housing censuses, Series M, No. 67 (E.70.VII.9). **5.3**

U.S. Census Bureau [1978] *The Current Population Survey: Design and Methodology*, Technical Paper 40. **4.7**, **6.2**, **6.4**, **7.1**

U.S. Dept of Commerce [1980], *Report on Statistical Uses of Administrative Records*. Washington: US Sup. of Documents. **6.1**

Velu, R and McInerney, M [1985] A note on statistical methods adjusting for intraclass correlation, *Biometrics*, **41**, 533–38. **7.1**

Verma, V, Scott, C, and O'Muircheartaigh, C [1980] Sample designs and sampling errors for the World Fertility Survey, *JRSS* (A), **143**, 431–473. **2.2**, **2.6**, **6.3**, **7.1**

Wagner, C H [1982] Simpson's paradox in real life, *The American Statistician*, **36**, 46–48. **4.2**

Waksberg, J [1968] The role of sampling in population censuses: its effect on timeliness and accuracy, *Demography*, **3**, 362–373. **5.2**

Wall, W D and Williams, H L [1972] *Longitudinal Studies and the Social Sciences*. London: Heinemann. **6.1**

Warner, S L [1971] The linear randomized response model, *JASA*, **66**, 884–88. **7.2**

Webb, E J, Campbell, D T, and Schwartz, R D, Seechrest L [1966] *Unobtrusive Measures.* Chicago: Rand McNally. **6.1**

Weller R H and Bouvier, L F [1981] *Population: Demography and Policy*, New York: St. Martin's Press. **3.7**

Winer, B J [1962] *Statistical Principles in Experimental Designs.* New York: McGraw-Hill. **1.3**

Wold, H [1956] Causal inference from observational data, *JRSS* (A), **119**, 28–61. **3.2**

Wolter, K M [1979] Composite estimation in finite populations, *JASA*, 1974, 604–613. **7.1**

Wright, C R [1967] Evaluation research, in the *International Encyclopedia of the Social Sciences.* New York: Macmillan. **3.7**

Wright, T and Tsao, H J [1983] A frame on frames: an annotated bibliography in T. Wright (ed.), *Statistical Methods and the Improvement of Data Quality.* New York: Academic Press. **2.1**

Wylie, L [1964] *Village in the Vaucluse.* New York: Harper Colophon. **3.1**

Yates, F [1935] Some examples of biased sampling, *Annals of Eugenics*, **6**, 202–13. **1.4**

Yates, F [1981] *Sampling Methods for Censuses and Surveys*, 4th ed. London: Griffin. The 1949 edition was the first modern book on sampling. **1.4**, **2.1**, **2.5**, **4.4**

Yule, G U and Kendall, M G [1965] *Introduction to the Theory of Statistics.* London: Griffin. **4.2**, **4.5**, **7.4**

Zarkovich, S S [1963] *Quality of Statistical Data.* Rome: FAO. **6.1**

Zelen, M [1969] Play the winner rule and the controlled clinical trial, *JASA*, **64**, 131–146. **4.3**

Index

The numbers following the entries refer to section numbers.

Applied Probability and Statistics (Continued)

JUDGE, HILL, GRIFFITHS, LÜTKEPOHL and LEE • Introduction to the Theory and Practice of Econometrics
JUDGE, GRIFFITHS, HILL, LÜTKEPOHL and LEE • The Theory and Practice of Econometrics, *Second Edition*
KALBFLEISCH and PRENTICE • The Statistical Analysis of Failure Time Data
KISH • Statistical Design for Research
KISH • Survey Sampling
KUH, NEESE, and HOLLINGER • Structural Sensitivity in Econometric Models
KEENEY and RAIFFA • Decisions with Multiple Objectives
LAWLESS • Statistical Models and Methods for Lifetime Data
LEAMER • Specification Searches: Ad Hoc Inference with Nonexperimental Data
LEBART, MORINEAU, and WARWICK • Multivariate Descriptive Statistical Analysis: Correspondence Analysis and Related Techniques for Large Matrices
LINHART and ZUCCHINI • Model Selection
LITTLE and RUBIN • Statistical Analysis with Missing Data
McNEIL • Interactive Data Analysis
MAINDONALD • Statistical Computation
MANN, SCHAFER and SINGPURWALLA • Methods for Statistical Analysis of Reliability and Life Data
MARTZ and WALLER • Bayesian Reliability Analysis
MIKÉ and STANLEY • Statistics in Medical Research: Methods and Issues with Applications in Cancer Research
MILLER • Beyond ANOVA, Basics of Applied Statistics
MILLER • Survival Analysis
MILLER, EFRON, BROWN, and MOSES • Biostatistics Casebook
MONTGOMERY and PECK • Introduction to Linear Regression Analysis
NELSON • Applied Life Data Analysis
OSBORNE • Finite Algorithms in Optimization and Data Analysis
OTNES and ENOCHSON • Applied Time Series Analysis: Volume I, Basic Techniques
OTNES and ENOCHSON • Digital Time Series Analysis
PANKRATZ • Forecasting with Univariate Box-Jenkins Models: Concepts and Cases
PIELOU • Interpretation of Ecological Data: A Primer on Classification and Ordination
PLATEK, RAO, SARNDAL and SINGH • Small Area Statistics: An International Symposium
POLLOCK • The Algebra of Econometrics
PRENTER • Splines and Variational Methods
RAO and MITRA • Generalized Inverse of Matrices and Its Applications
RÉNYI • A Diary on Information Theory
RIPLEY • Spatial Statistics
RIPLEY • Stochastic Simulation
RUBIN • Multiple Imputation for Nonresponse in Surveys
RUBINSTEIN • Monte Carlo Optimization, Simulation, and Sensitivity of Queueing Networks
SCHUSS • Theory and Applications of Stochastic Differential Equations
SEAL • Survival Probabilities: The Goal of Risk Theory
SEARLE • Linear Models
SEARLE • Matrix Algebra Useful for Statistics
SPRINGER • The Algebra of Random Variables
STEUER • Multiple Criteria Optimization
STOYAN • Comparison Methods for Queues and Other Stochastic Models
TIJMS • Stochastic Modeling and Analysis: A Computational Approach
TITTERINGTON, SMITH, and MAKOV • Statistical Analysis of Finite Mixture Distributions
UPTON • The Analysis of Cross-Tabulated Data
UPTON and FINGLETON • Spatial Data Analysis by Example, Volume I: Point Pattern and Quantitative Data

(*continued from front*)